MEDICAL
INTELLIGENCE
UNIT

LOCAL LIPOSOME DRUG DELIVERY

MEDICAL
INTELLIGENCE
UNIT

LOCAL LIPOSOME DRUG DELIVERY:
AN OVERLOOKED APPLICATION

Carl I. Price, M.D.
Emory University
Atlanta, Georgia

Jureta Horton, Ph. D.
University of Texas Southwestern Medical Center
Dallas, Texas

R.G. LANDES COMPANY
AUSTIN

MEDICAL INTELLIGENCE UNIT

LOCAL LIPOSOME DRUG DELIVERY:
AN OVERLOOKED APPLICATION

R.G. LANDES COMPANY
Austin / Georgetown

CRC Press is the exclusive worldwide distributor of publications of the Medical Intelligence Unit.
CRC Press, 2000 Corporate Blvd., NW, Boca Raton, FL 33431. Phone: 407/994-0555.

Production Manager: Judith Kemper
Copy Editor: Constance Kerkaporta

Please address all inquiries to the Publisher:
R.G. Landes Company
909 Pine Street
Georgetown, TX 78626
or
P.O. Box 4858
Austin, TX 78765
Phone: 512/ 863-7762
FAX: 512/ 863-0081

ISBN 1-879702-12-6
CATALOG # LN0212

FOREWORD

Why is a surgeon interested in liposomes? One reason is the long time interest of surgeons in septic shock. This has led to a continuing search for more effective methods of treating bacterial infections, a major source of sepsis. In addition, surgeons are often confronted with infections in a localized anatomic area. Surgical drainage along with systemic antibiotics are commonly used to treat these localized infections with frequent failure due to inadequate drainage or inability of the antibiotic to penetrate the infected site. When recalcitrant to conventional therapy, these infections result in treatment failures and progressive sepsis. Furthermore, in surgery, despite widespread antibiotic prophylaxis, serious postoperative iatrogenic infections still occur. Prophylactic and treatment failures obviously increase patient morbidity, mortality and ultimately the cost of medical care in both human misery and monetary terms. A method of enhancing antimicrobial effectiveness in the surgical settings of surgical prophylaxis and in the treatment of localized infectious processes has immediate widespread utility. Liposomes offer one solution to these and other surgical problems.

There are various reasons why liposomes are appealing as local drug delivery vehicles in surgical settings. In surgery, it is common to have a body cavity or organ exposed and to directly apply single or combination antibiotics. This practice of direct antimicrobial application, while common both as a method of treatment and as prophylaxis against infections, is not wholly supported by the literature.[1-3] A reason for the conflicting opinion is the rapid absorption of a locally applied agent from an application site which may lead to variable drug activity. Liposomal delivery presents one means of inhibiting rapid absorption of drug from a local site while retaining drug efficacy. Moreover, the sequestration and clearance of liposomes from local sites by the RES system may even enhance the efficacy above that resulting from the increased local concentrations alone. These properties also indicate that liposomes may find effective use in cancer therapeutics and with various wound modulators.

From a review of the literature prior to beginning our liposome work, it appeared that the early laboratory excitement surrounding liposome use could not be translated into clinically useful applications.[4] Liposomes were, in many ways, a technology looking for an application. On the other hand, the various surgical situations like those mentioned above seemed to call for a new delivery method, like liposomes, that could enhance the local use of different agents.

ACKNOWLEDGEMENTS

I would like to acknowledge several people who were instrumental in the investigations of local liposomal delivery: first the co-author, Jureta Horton, without whose encouragement, help, patience and suggestions, the quality and quantity of our work would be much diminished. The next is Claude Organ, M.D. who as a surgeon-educator constantly challenged young housestaff to justify the dogma which we daily practiced. This inquisitive approach has continued to permeate investigations. Kathy Sigler Price, Pharm. D. has been invaluable as an editor and confidant. Also, without the fertile environment provided by G. Rainey Williams M.D., Maurice J. Jurkiewicz, M.D. and John Bostwick III, M.D. none of this work would have come to fruition. Finally, I would like to acknowledge the support of Charles R. Baxter, M.D. who is one of the visionaries in surgery. He had the insight to realize the potential of liposome technology, was excited about it, and allowed the leeway and support to do with it what I thought appropriate.

CONTENTS

CHAPTER 1

INTRODUCTION

Liposomes have been investigated as drug delivery vehicles since their discovery by Bangham[5] and Horne in the early 1960s. While there has been considerable laboratory investigation into the potential uses of this technology, few useful products have resulted. Reasons for the lack of tangible benefits from the applied liposome technology range from regulatory obstacles to the complexity of lipid biochemistry in biological systems.[6] Also, much of the initial enthusiasm and subsequent investigation have focused on the use of liposomes as carriers of agents to target specific tissue or organs with systemic administration. Complex efforts to formulate liposome so as to effect more specific targeting have included a variety of methods ranging from the coating of liposomes with polyethylene glycol to monoclonal antibodies. Further hampering the efforts to use liposomes effectively is the variable behavior of different liposome formulations in biologic systems and the ability of incorporated compounds to potentially alter the liposomal membrane, thus modifying the liposome characteristics and changing biologic behavior. This leads not only to problems with carrier drug development but also to problems in regulatory approval for a carrier alone. Finally, with the prospect of lengthy and costly approval for every agent incorporated into liposomes there is some disincentive to develop liposome technology by the pharmaceutical industry.

There are other complexities that complicate systemic administration of drugs with liposome systems. The physical size of some formulations can prevent the easy egress through capillary membranes of some tissues, keeping liposomal pharmaceuticals in the intravascular space and out of targeted tissues.[7,8] Also preparations which target the RES system may have variable in vivo characteristics because host specific antigens can easily become embedded in the lipid membrane, producing an inconsistent antigenic profile and therefore inconsistent RES recognition.[8] Complex refinements in liposomal construction required to solve these problems can make their general clinical application cost prohibitive. A more direct approach which seems to obviate many of the complexities and problems of systemic liposomal administration is to apply liposome entrapped agent directly to the site where the therapeutic agent is needed and then use the properties of liposome to favorably modify its characteristics. This has been attempted with some chemotherapeutic agents[10] and certain antimicrobial agents[4,9] however, these experimental investigations have tended to have limited direct clinical application.

In surgery, there is already widespread use of local antibiotics including topical application to open wounds, topical application in burn wound prophylaxis, application to subcutaneous tissues and the irrigation of body cavities with antibiotic solutions (most often the peritoneum). The potential benefits of liposomal antibiotic delivery could result in the widespread use of this technology in these and other surgical settings.

This monograph will briefly discuss the biochemistry of liposomes and then concentrate on the applications and clinical possibilities of local antimicrobial liposome use as explored in our lab. Other potential applications for local liposomal drug delivery in the areas of cancer chemotherapeutics, immunomodulation, free radical scavenging, and gene therapy will be discussed to hopefully enlighten and stimulate the reader to the overlooked possibilities of this exciting technology.

REFERENCES

1. Lord JW, LaRaja RD, Daliana M, Gordon MT. Prophylactic antibiotic irrigation in gastric, biliary, and colon surgery. Amer J Surg 1983; 145:209-212.
2. Liebhoff AR, Soroff HS. The treatment of generalized peritonitis by closed peritoneal lavage: A critical review of the literature. Arch Surg 1987; 22:1005-1010.
3. Raahave D, Hesselfeldt P, Pederson T, Zachariassen A, Kann D, Hansen OH. No effect of topical ampicillin prophylaxis in elective operations of the colon or rectum. Surg Gynecol Obstet 1989; 168:112-114.
4. Roerdink FH et al. Therapeutic utility of liposomes. drug delivery systems: Fundamentals and technics. In: Ed Johnson P and Lloyd-Jones JG, eds. Ellis Horwood, 1987:66-80.
5. Bangham AD. Liposomes in biological systems. In: Gregoriadis G, Allison AC, eds. Development of the Lipisome Concept. John Wiley and Sons, Ltd, 1980:1-24.
6. Ostro MJ. A medical wonder that isn't - yet. The New York Times 1990:F12.
7. Ostro MJ. Liposomes. Sci Amer 1987; 256:102-111.
8. Juliano RL. Liposomes as drug carriers in the therapy of infectious diseases. Horizons in Biochemistry and Biophysics 1989; 9:249-279.
9. Peyman GA, Charles HC, Liu KR, Khoobehi B, Niesman M. Intravitreal liposome-encapsulated drugs: A preliminary human report. International Ophthalmology 1988; 12:175-182.
10. Delgado G et al. A Phase I/II study of intraperitoneally administered doxorubicin entrapped in cardiolipin liposomes in patients with ovarian cancer. Amer J Obstet Gynecol 1989; 160(4):812-819.

CHAPTER 2

BIOCHEMISTRY OF LIPOSOMES AND LIPOSOME FORMULATION

Liposomes are synthetic lipid vesicles consisting of phospholipids; the lipid bilayer, similar in structure and composition to biological membranes, encloses an aqueous volume. Various agents can be entrapped in the vesicular compartment, providing an effective means of transporting agents. While the recent use of liposomes in cosmetics has changed the concept of skin care, liposomes likely have a major role in the pharmaceutical industry as a delivery vehicle for drugs and genetic material.

STRUCTURE AND PROPERTIES

The membranes of liposomes are most frequently phospholipids, routinely containing two long chain fatty acids which are linked via a three-carbon glycerol with a phosphoryl choline entity; numerous fatty acids, usually one saturated and one unsaturated and forming the hydrophobic portion of the molecule, occupy positions one and two of the glycerol bridge while the polar hydrophilic group is phosphocholine. Though insoluble in water, the phospholipid (phosphatidyl choline molecules) form a bilayer arrangement in aqueous media, with the hydrophobic portion protected from water molecules by orienting the

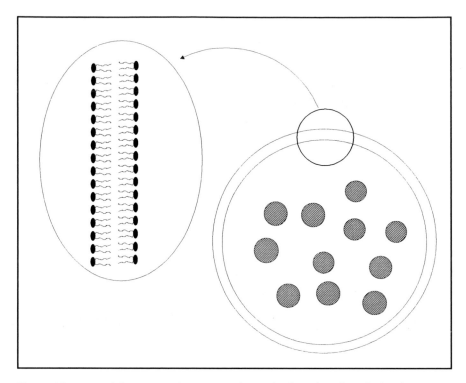

Fig. 1. Liposome bilayer membranes are formed when the phospholipids arrange themselves such that the hydrophilic heads are oriented towards the aqueous phase and the hydrophobic tails away from the aqueous phase.

water insoluble fragment in an inward direction; the hydrophilic or water soluble portion is oriented outward toward the aqueous phase. (Fig.1)

The bilayer arrangement of phosphatidyl choline contrasts with the micellar structure of lysolecithin and amphipathic molecules. The bilayer arrangement of phosphatidyl choline, because of instability, tends to fold producing vesicles or closed compartments characterized by semipermeable membrane which effectively restrict movement of polar and high molecular weight compounds. Therefore, incorporation or entrapment of larger polar molecules in the aqueous interior of the liposome prevents diffusion across the membrane barrier or allows diffusion across the membrane barrier very slowly, while small molecules such as water move rapidly and freely across the membrane. Movement of sodium and potassium and metal ions across the liposome membrane is dependent upon the saturation or unsaturation of the fatty acid components of the liposomes. Incorporation of lipophilic agents into the molecule (1-10% by weight), while altering membrane fluidity and increasing the outer layer surface, does not disrupt the bilayer structure of the liposome. In addition to the incorporation of lipophilic compounds into the liposomal membrane and entrapment of water soluble molecules in the intravesicular aqueous compartment, numerous agents such as amphipathic compounds can be incorporated into the phospholipid bilayer. Incorporation of agents into phospholipids is physical in nature, preserving the chemical nature of the incorporated compound. Phosphotidyl choline can be extracted from a number of sources, in our laboratory from soybean. The formulation of liposomes from natural phospholipids is based on their inertness and their low cost.[1-4]

The liposomes used in studies from our laboratory are a proprietary formulation of Fountain Pharmaceuticals, Largo, FL. The liposomes are constructed in a solvent dilution technic using mainly soy phosphatides. The resultant liposomes are fairly uniform at 0.34 microns, have one to few lipid bilayers (unilamellar to oligolamellar) and

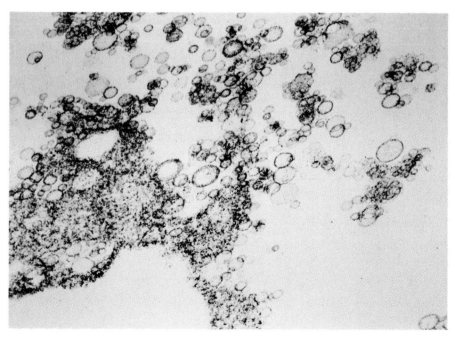

Fig. 2. Electron photomicrograph of the liposome preparation used in our studies.

a net negative charge. (Fig 2) The rate of encapsulation is regularly over 90%. One of the striking advantages of this particular formulation is that it is purported to be very cost effective and easy to scale up to commercial scale.[5]

Liposomes vary in size and shape with multilamellar vesicles 100-1000 nm in size and composed of multiple (usually greater than five) concentric lamellae; unilamellar vesicles are 15 to 25 nm in diameter and consist of a single bilayer membrane; large unilamellar vesicles consist of a single bilayer with a diameter of 1000 nm while intermediate unilamellar vesicles have a diameter of 100 nm. The volume entrapped within the unilamellar liposome varies directly with the liposome size while the entrapped volume within the multilamellar liposomes varies with the quantity of lipid and the configuration of the lamellae (equal spacing of bilayers versus unequal arrangement of the lipid bilayers). In preparing liposomes, incorporation of hydrocarbon chains greater than 14 carbon atoms in length and characterized by single bonds produce gel-like liposomes which are more stable with less leakiness of incorporated drug. Formulation of liposomes with specific attention to the chemical characteristics of the lipids, the size of the liposomes, as well as the number of lipid layers can produce liposomes as effective vehicles for specific drug delivery. In this way, liposomes can be formulated to target specific organs and to release their contents over varying time periods; in addition, other components can be incorporated into the liposome membrane to produce specific interaction with biological systems or to modify the action of the liposome.[1,6-8]

LIPOSOMES FOR DRUG DELIVERY

In using liposomes as a drug delivery system, the drug can be incorporated into the aqueous media or bound to the liposome membrane by electrostatic interaction; specific tailoring of the liposome enables the investigator to trigger release of the liposome contents at a specific temperature, rapidly, or gradually over a longer period of time. Proteins can be incorporated into the liposome membrane to promote binding with specific cell receptors; liposomes can also be tailored with regard to size, formulating liposomes of intermediate size to ensure retention within the vascular compartment after intravenous administration. While most liposomes are retained in the vascular compartment after intravenous injection due to the inability to traverse the capillary, the presence of sepsis and/or inflammation produce leaky membranes, allowing transcapillary movement of liposomes.

Therefore, in disease states characterized by a leaky membrane syndrome, transcapillary movement of liposomes can occur.[3] After IV administration, liposomes which stay within the intravascular compartment can interact with opsonins or plasma proteins which adhere to the liposome outer surface, rendering the liposome more susceptible to phagocytosis by cells of the reticuloendothelial system; degradation of the liposome then occurs within the reticuloendothelial system. While intermediate size liposomes are retained within the vascular compartment, small liposomes can exit the vascular compartment within a peripheral organ, achieving direct contact between liposomes and individual tissue cells. Regardless of liposome formulation with regard to physical size, chemical properties and membrane composition, most liposomes ultimately react with the reticuloendothelial system and are sequestered in the liver, lung, spleen, lymph nodes, and bone marrow. Liposomes are engulfed by monocytes/macrophages by endocytosis, and incorporated into lysosomes, degraded by phospholipases, releasing free drug/agent into the intracellular space.[9,10]

Therefore, the usefulness of liposomes in the treatment of liver or lung disease or specific macrophage diseases such as leishmaniasis is obvious. The role of macrophages in clearing diseased cells, invading bacteria, and foreign particles has contributed to the major development of liposomes as delivery vehicles to treat diseases of the reticuloendothelial system. Treatment of leishmaniasis, a parasitic disease estimated to infect over 100 million people worldwide, has effectively utilized liposome encapsulated antimony-containing compound; while these drugs are toxic in a free state, liposome encapsulation specifically directs the drug to macrophages, reducing toxicity and increasing therapeutic effectiveness.[11,12]

Liposomal entrapped drugs have been particularly effective against some tumors, concentrating tumor-specific drug within the tumor while lessening the systemic toxic side effects. In addition, incorporation of specific components into liposome membranes can effectively stimulate immune reaction by antigenic presentation to macrophages. Subcutaneous and intramuscular administration of liposomes result in lymphatic uptake while orally administered liposomes are degraded principally in the gastrointestinal tract with limited uptake by gastrointestinal lymphatics.[13,14]

Another successful application of liposome technology has been the treatment of fungal infections such as *Candida albicans.* These infections are particularly prevalent in patients who are immunocompromised as a result of chemotherapy or the AIDS infection. Amphotericin B treatment, while effective, has been associated with kidney and central nervous system toxicity; liposome encapsulation of amphotericin has eliminated these toxic side effects.[2]

Since liposomes are ingested by macrophages and readily cleared from the vascular compartment, several attempts have been directed toward decreasing liposome-mediated activation of the immune system and subsequent clearance by the macrophages; this effort would significantly increase the circulating half-life of the liposomes. In this

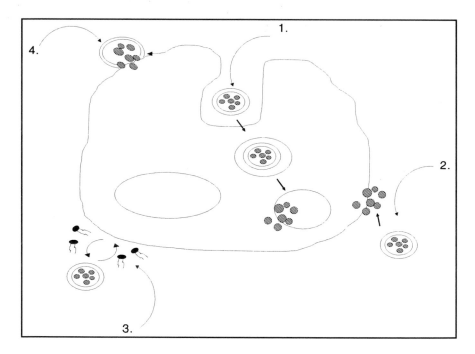

Fig. 3. Methods of liposome drug delivery. 1) Endocytosis. Liposomes are taken up by the cell and then either deliver drug into the cytosol or are degraded in lysosymes and thus release drug. 2) Diffusion or Adsorption. Liposome encapsulated drug simply diffuses down a concentration gradient. 3) Lipid exchange. Phospholipid and drug are exchanged with phospholipids that make up the cell membrane. 4) Fusion. Liposomes fuse with the cell membrane releasing drug into the cytosol.

regard, coating liposomal surfaces with polyethylene glycol chains, a coating similar to the sialic acid coating of a red blood cell, a normal protective mechanism by which the red blood cell avoids destruction by the immune system, has successfully increased the half-life of liposomes in the vascular compartment. It is likely that the polyethylene glycol coating on liposomes increases transit time in the vascular compartment by interfering with adherence of opsonins to the liposomal surface. Opsonins become embedded in the outer layer of the liposome, identifying the liposome as foreign material and promoting ingestion by macrophages. A protective polyethylene glycol coating on the liposomes likely prevents opsonin adherence, increasing vascular persistence of the liposomes.[10,15]

The mechanisms by which liposomes deliver their contents remains unclear; some studies have shown that contact of liposomes with cells trigger release of content secondary to alterations in liposomal membrane characteristics, eliminating the need for ingestion or degradation of the liposome. Another mechanism of liposome-drug release is the specific binding of liposomes to cell surface receptors; the adsorption process may be followed by phagocytosis or pinocytosis of the liposomes. Alternatively, physical approximation of liposomes and cell membranes may produce fusion of the two components and subsequent mixing of the liposome aqueous content with the intracellular matrix. (Fig.3) Liposomes which have been constructed so as to be temperature or pH sensitive, are rapidly degraded at temperatures greater than 40^0C or at a pH of 5.0.[1,16]

Topical application of medications such as creams or gels to the skin surface have provided another means of drug delivery; however, rapid absorption of these topically applied drugs frequently results in toxic systemic concentrations. Encapsulation of topical agents into liposomes can increase local drug concentrations, increase the time of selective delivery at the application site, and finally avoid the possible toxic side effects of achieving high systemic drug levels. Topical application of liposome containing agents has been proposed for hormonal therapy, delivery of anti-inflammatory drugs and antibiotics, as well as delivery of moisture and lipid agents for cosmetic purposes.[1,2,17,18]

Finally, liposomal transport of genes or proteins directly into individual cells, sometimes referred to as "lipofection" may contribute significantly to genetic engineering research. This type of gene replacement for treatment of numerous diseases requires a protective transport system for the genetic material to ensure cell recognition and incorporation of the genetic material into either the cellular cytoplasm or the intracellular lysosomes. After deposition of the genetic material in the cellular cytoplasm, the DNA or RNA may fuse with the cell membrane or be incorporated into the nuclear matrix.[2]

The use of biological membranes as delivery vehicles for a vast array of materials appears to have application in botany, cell biology, genetics, as well as clinical medicine. Development and application of liposomal technology in the pharmaceutical and cosmetic industry has just begun.

REFERENCES

1. New RRC. Introduction. In: New, RRC, ed. Liposomes a practical approach. Oxford: Oxford University Press, 1990:chapt 1.
2. Lasic D. Liposomes. American Scientist 1992; 80:20-31.
3. Gregoriadis G. The physiology of the liposome. News Physiol Sci (NIPS) 1989; 4:146-151.
4. Batzri S, Korn ED. Simple bilayer liposomes prepared without sonication. Biochimica et Biophysica Acta 1973; 298:1015-9.
5. Fountain MW. Fountain Pharmacueticals. Personal communication.
6. Woodle MC, Papahajopoulos. Liposome preparation and size characterization. Methods in Enzymology 1989; 171:193-217.
7. Papahadjopoulos D, Miller N. Phospholipid model membranes: Structural characterization of hydrated liquid crystals. Biochimica et Biophysica Acta 1967; 135:624-638.
8. Bangham AD, Hill MW, Miller NGA. Preparation and use of liposomes as models of biological membranes. Methods Membr Biol 1974; 1:1-68.
9. Lasic DD. A general model of vesicle formation. J Theoretical Biol 1987; 124:35-41.
10. On the formation of membranes. Nature 1991; 351:163.
11. New RRC, Chance ML, Thomas SC, Peters W. Anti-leishmanial activity of antimonial entrapped in liposomes. Nature 1978; 272:55-56.
12. Alving CR, Steck EA, Chapman WL, Waits VB, Hendricks LD, Swartz GM, Hanson WL. Therapy of leishmaniasis: Superior efficacies of liposome-encapsulated drugs. Proc of the Natl Acad Sci 1978; 75(6):2959-2963.
13. Gregoriadis G. Liposomes as carriers of drugs. 1988 Chichester; J Wiley and Sons.
14. Chiang CM, Weiner N. Gastrointestinal uptake of liposomes. In vivo studies. Int J Pharm 1987; 40:143-150.
15. Papahadjopoulos D, Allen TM, Gabizon A et al. Sterically stabilized liposomes: pronounced improvements in blood clearance, tissue distribution and therapeutic index of encapsulated drugs against implanted tumore. Proc Natl Acad Sci 1991; 88:11460-11464.

16. Leserman LD, Barbet J, Kourilsky F, Weinstein JN. Targeting to cells of fluorescent liposomes covalently coupled with monoclonal antibody or protein A. Nature1980; 288(5791):602-4.

17. Rowe TC, Mezei M, Hilchie J. Treatment of hirsutism with liposomal progesterone. The Prostate 1984; 5:346-347.

18. Patel HM. Liposomes as a controlled-release system. Biochem Soc Trans 1985; 13:513-516.

CHAPTER 3

Topical Liposome Delivery

S mall superficial wounds are common in clinical medicine. Most of these wounds respond to local treatment consisting of debridement and frequent dressing changes. Single-species bacterial proliferation in these wounds requires additional treatment, usually with topical antimicrobials, to decrease bacterial concentration and to promote healing. Intermittent topical application of antimicrobials is often inefficient because of the rapid absorption of the drug from the wound; however, frequent or continuous application may be effective.

Since liposomes have been used as drug delivery vehicles to change the pharmacokinetics and pharmacodynamics of administered antineoplastic and antimicrobial agents[1-4] our initial studies focused on the question of whether liposome encapsulated antibiotics would result in adequate therapeutic effects when applied topically and compared to topical free antibiotic application. In particular, the antibacterial effectiveness of a single dose of liposome encapsulated antibiotic, delivered via incorporation into a polyurethane sponge, was explored. As stated earlier, most clinical and experimental studies of liposomes have involved intravenous administration. With intravenous administration the liposomal use has been directed primarily toward drug delivery where toxicity is a problem or to increase the drug concentration in sites where directed action is desirable, such as the reticuloendothelial system.[5] Fountain and colleagues showed that RES targeting was possible and desirable in a study on *Brucella abortus* infections. In this study pluralamellar vesicles containing aminoglycosides eradicated the *B. abortus* infection from the RES cells of infected mice while similar dosages of free drug did not.[6] Sells et al have also showed the ability of liposome encapsulated doxorubicin to decrease the cardiotoxicity of this drug.[5]

Local liposome delivery of pharmacologic agents has been explored in a limited fashion. Delgado and associates have shown that intraperitoneal application of liposome-encapsulated doxorubicin in human subjects with advanced ovarian cancer allowed for significantly higher dosages of drug to be given without the side effects that would be expected with free drug.[7] Weiner and his group have explored the release of agents from a liposome-collagen gel matrix.[8] Hong and Mayhew have showed complete response of experimental central nervous system L1210 monoclonal leukemia to intracranially applied 1-*B*-D arabinofuranosylcytosine encapsulated in liposomes. Free drug applied similarly did not have as dramatic a response.[9] Jackson[16] demonstrated the local persistence of intramuscularly injected liposome encapsulated inulin compared to unencapsulated inulin, which was rapidly absorbed. These studies indicate that liposome encapsulation can favorably modify entrapped agents applied locally.

An early study in our laboratory was designed to explore the potential of administering a single dose of topically applied antimicrobials via liposomal carriers in an experimental model of superficial infection.[10] The effect of liposome entrapped antimicrobials on bacterial colony counts of infected animals was compared with bacterial counts of animals treated with free antimicrobial agents as well as with bacterial counts measured in untreated animals.

In these, our initial studies, a nonlethal model of superficial infection was initiated in 155 adult Sprague-Dawley rats by injection of a clinical isolate of *Pseudomonas aeruginosa* superficially under the fascia of the paraspinus muscle. All wounds were dressed with N-Terface®, a nonadherent wound material, and covered with Kontor® sponge, an open-cell polyurethane sponge containing either normal saline (group I), free tobramycin (groups III and V), liposome-entrapped tobramycin (groups II and IV), silver sulfadiazine (group VI), or liposome-entrapped silver sulfadiazine (group VII). Free drugs were administered diluted in 3 ml of normal saline every 12 hours. Liposome

Table 1. Natural Log of the Number of Colony Forming Units of Pseudomonas per Gram of Tissue

	24hrs	48hrs	72hrs
untreated (group I)	15.66±0.26 (n=11)	16.32±0.28 (n=10)	15.77±0.23 (n=8)
tobramycin 0.24mg/dose (group II)	15.93±0.63 (n=10)	10.46±0.81 (n=10)*	12.16±0.97 (n=10)*
LET 1.7mg (group III)	14.49±0.50 (n=9)	12.49±0.73 (n=8)*	12.41±0.90 (n=8)*
tobramycin 2.4mg/dose (groupIV)	9.30±0.56 (n=10)*	7.53±0.66 (n=10)*	7.84±1.01 (n=10)*
LET 17mg (group V)	10.26±0.59 (n=10)*	8.46±0.92 (n=9)*	9.04±0.63 (n=10)*
SSD 11mg/dose (group VI)	*****	12.75±0.39 (n=10)*	*****
LESSD 60mg (group VII)	*****	12.56±0.37 (n+9)*	*****

LET - Liposome entrapped tobramycin
LESSD - liposome entrapped silver sulfadiazine
Expressed as the mean + S.E.M.
* indicates significant difference from untreated animals.
$P < 0.05$ (ANOVA).

drugs were incorporated into the sponge matrix and were, therefore, given as a single dose. The amount of drug incorporated into the liposome group was equivalent to the total amount of free drug that would be given over three days. The sponges containing liposomes were moistened every 12 hours with 3 ml of normal saline. Animals were sacrificed at 24, 48 and 72 hours. At sacrifice, the superficial muscle in the area of the wound was excised to determine the number of colony-forming units per gram of tissue (Table 1).

Liposome-entrapped tobramycin was given in two different doses 1.7 and 17mg. Liposome-entrapped silver sulfadiazine was given in a

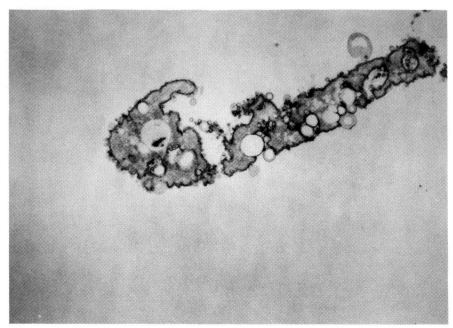

Fig. 1. Electron photomicrograph showing the liposomes forming at the border of the liposome sponge matrix. 15000x.

60mg dose. The liposome antibiotic solution was incorporated into a polyurethane sponge. Upon being exposed to aqueous medium, the liposomes were released. As revealed by electron microscopy most liposomes formed mostly along the junction of the sponge border. (Fig. 1)

In this study liposome delivery of antimicrobials resulted, after one application, in an effect on tissue bacterial counts that required multiple doses of free drug. While numerous studies have described the use of liposomes to administer pharmacologic agents, as previously mentioned, most studies have focused on the intravenous administration of liposome agents. The property of liposomes most favorable for local delivery of drugs is that they tend to stay in the intravascular or extravascular space where they are applied.[1,11,12] This property appears to be advantages when treating localized infections as the one described in this experimental model. The persistence of the liposome and its entrapped antibiotic as well presumed RES clearance from the local site may contribute to some enhanced efficacy although none was demonstrated in this initial study. Our initial data demonstrated that the liposome antibiotic is at least as efficacious as free drug in decreasing the bacterial counts in topically infected animals. More significant is the fact that only a single dose of the liposome antibiotic was given to achieve the same effect on bacterial counts as did multiple doses of free drug. Also significant is the fact that the liposome preparation combined with the antibiotic could be applied to the polyurethane sponge, dehydrated, and then be reactivated by an application of saline with antibacterial activity intact.

In vitro studies have shown that liposomes incorporated into the sponge matrix are released over a period of weeks in a linear fashion.[13] Subtle differences in efficacy between conventional free antibiotics and liposome-encapsulated antibiotic may not be apparent at higher doses of drugs, since at high antibiotic concentrations an overwhelming bacteriocidal effect may be present, masking differential responses of liposome carried drug versus conventional free drug. At lower dosage where incomplete bacteriocidal effects are present, subtle differences are obvious. In the low dose tobramycin groups, similar bacterial counts were seen at each time period regardless of whether free or liposome drug were used. In contrast to the work of Fountain with *B. abortus* and Lopez-Berestein with candida, initial inspection of our data indicate no enhanced effectiveness with liposome delivery of antibiotics.[6,14] Further examination of this apparent lack of advantage with liposome drug led to interesting speculation. Although the total amount of drug in the liposome-sponge matrix was equal to the amount of free drug given over a 72-hr period, encapsulated drug was released from the sponge matrix over a period of weeks, assuming that the liposome drug release was similar to the liposome release alone. Thus only a fraction of the total encapsulated dose was actually released during the 72-hr study period, resulting in smaller amount of drug in the wound on a daily basis in the liposome groups. This speculation suggested that since the effect on bacterial counts was similar between the liposome and free drug groups, the advantage of liposome delivery was that an equal bactericidal effect was accomplished with an effectively lower dose of drug. These initial encouraging results led to our further investigations with local liposome antibiotic administration in conventional bacterial infections.

Topical application of liposomal agents has widespread application. In the cosmetic industry various products are currently available that use liposomes to prolong and enhance the action of skin care agents. Extrapolation of our liposome sponge model to the clinical setting suggest that the liposome antibiotic placed in the sponge matrix could potentially be used in contaminated superficial wounds to decrease bacterial counts with less frequent dressing changes and medicine application. Possibly higher total medication quantity could be used with less chance of side effects or toxicity. A specific application could be the treatment of small wounds with a band-aid type of a dressing. The liposome antibiotic would keep the bacterial counts down allowing for more rapid healing. This clinical application could also expand to include treatment of postoperative surgical wounds with antimicrobials to prevent infections. Moreover as our understanding of wound healing becomes more sophisticated wound modulating agents could be applied in this manner. This could include the various wound healing factors currently under investigation and free radical scavengers all of which tend to have vary short tissue half-lives. Even local anesthetics encapsulated in liposomes, which may be be more effective in the relief of pinprick pain than a eutectic mixture of local anesthetics (EMLA), by virtue of increased lipophilicity, could be applied in this manner to relieve postoperative or posttraumatic incisional pain.[15]

Liposome delivery could also be valuable in situations where topical agents result in local irritation. Since liposome incorporation has the potential to decrease the local irritation caused by some agents, as demonstrated by the lack of phlebitis and venous sclerosis seen by Sells and associates when doxorubicin was encapsulated in liposomes, it may be possible to apply agents which are currently underutilized.[5] For example topical 5-FU is used to treat various skin disorders. Liposome encapsulation could reduce the local and systemic side effects of treatments such as this ultimately allowing for higher and more extended dosing regimes.

Topical administration of liposomal agents certainly has many useful medical applications. This may be the area of where liposomal agents are first widely used outside of the cosmetic industry.

References

1. Ostro MJ. Liposomes. Sci Amer 1987; 256:102-111.

2. Fountain MW, Dees C, Schulte RP. Intracellular killing of S aureus by liposomes containing aminoglycosides: A comparison of liposome surface charge and liposome entrapment of amikacin, tobramycin, gentamicin and kanamycin. Curr Microbiol 1981; 6:373-376.

3. Amber L. Liposome encapsulation enhances antibiotic activity. Infec Dis Today 1988; 2:26

4. Lopez-Berestein G, Fainstein V, Hopfer R et al. Liposomal amphotericin B for the treatment of systemic fungal infections in patients with cancer: a preliminary study. J Infect Dis 1985; 151:704-710.

5. Sells RA, Gilmore IT, Owen RR, New RRC, Stringer RE. Reduction in doxorubicin toxicity following liposomal delivery. Cancer Treat Rev 1987; 14:383-387.

6. Fountain MW, Weiss SJ, Fountain AG, Shen A, Lenk RP. Treatment of Brucella canis and Brucella abortus in vitro and in vivo by stable pleurilamellar vesicles encapsulated aminoglycoside. J Infect Dis 1985; 152:529-536.

7. Delgado G et al. A Phase I/II study of intraperitoneally administered doxorubicin entrapped in cardiolipin liposomes in patients with ovarian cancer. Amer J Obst Gynecol 1989; 160(4):812-819.

8. Weiner AL, Carpenter-Green SS, Soehngen EC, Lenk RP. Liposome-collagen gel matrix: A novel sustained drug delivery system. J Pharm. Sci 1985; 74:922.

9. Hong F, Mayhew E. Therapy of central nervous system leukemia in mice by liposome entrapped 1-B-D arabinofuransylcytosine. Cancer Res 1989; 49:5097-5107.

10. Price CI, Horton JW, Walker PB. Model of superficial soft tissue infection in the rat. FASEB J 1989; 3:A633.

11. Juliano RL. Liposomes as drug carriers in the therapy of infectious diseases. Horiz in Biochem and Biophy 1989; 9:249-279.

12. Hirano K, Hunt CA. Lymphatic transport of liposome-encapsulated agents: Effects of liposome size following intraperitoneal administration. J Pharm Sci 1985; 17:915-921.

13. Fountain MW. Personal communication.

14. Lopez-Berestein G. Liposomes as carriers of antifungal drugs. Ann NY Acad Sci 1988; 544:590-597.

15. McCafferty. In vivo assessment of percutaneous local anaesthetic preparations. Br J Anaes 1989; 62:17-21.

16. Jackson AJ. Intramuscular absorption and regional lymphatic uptake of liposome-entrapped inulin. Drug metab Disp 1981; 9:535-540.

CHAPTER 4

The Subcutaneous Application of Liposome Antibiotics: Prophylaxis and Treatment

Surgical wounds often have bacterial contamination that can result in devastating infectious complications. While prophylactic antibiotics can partially decrease the incidence of these infections, these antibiotics should have certain qualities to be effective.[1-4] These qualities include ease of administration, antimicrobial activity against expected pathogens and adequate duration of antimicrobial activity relative to the length of exposure to potential contaminants. In addition, the antibiotic should present little risk to the patient with regard to adverse reactions and side effects. Systemic administration of prophylactic antibiotics promotes wide distribution of drug but requires doses sufficient to ensure adequate drug concentrations at the site of potential contamination. In addition, the drug must be given before the start of a surgical procedure to ensure that adequate drug levels are achieved in the "at risk" site during the entire exposure period. The application of antibiotics directly to the surgical site offers the possibility of targeting only the tissues at risk for contamination while reducing exposure of other tissues to the potentially toxic effects of the drug. Local antibiotic use also results in local tissue concentrations that are transiently higher than those that can be otherwise obtained with parenteral antibiotics.[5,6]

While the use of local antibiotics in surgical wound prophylaxis remains controversial, topical prophylactic antimicrobials have been used quite effectively to prevent burn wound infections. In addition, local antimicrobial use in complicated biliary surgery has been shown to be equivalent to systemic prophylactic antibiotics.[5,7] Another effective application of local antibiotic use has included the direct application of vancomycin paste to the sternum during cardiac surgery; this local antibiotic use has been shown to decrease the incidence of sternal wound complications.[8] Recently ReMine and Organ have reported that continuous local antibiotic irrigation is of benefit in the treatment of heavily contaminated wounds.[9] In general, however, the efficacy of local antibiotic prophylaxis has been variable, likely due to the transient nature of tissue antibiotic concentrations after local application. Because the locally applied drug is quickly absorbed, the drug pharmacology becomes similar to that achieved with systemic administration and any advantages of local administration (higher local drug concentrations and decreased systemic toxicity) are quickly lost.

In the early days of the antibiotic era, local treatment of infected wounds with sulfonamides was widespread. As early as 1948, Florey and Williams showed that locally applied penicillin was effective in the treatment of localized hand infections.[10] Currently, the use of local antibiotics in the treatment of infections is limited, likely due to the fact that there has been a proliferation of potent broad-spectrum antibiotics and the concern that rapid absorption of locally applied antibiotics limits the potential advantages of this administration route. Orthopedists have addressed this concern of rapid absorption by using antibiotic impregnated methylmethacrylate beads to increase local concentrations of antibiotic and decrease the absorption.[11] The necessity of this was born of the fact that even newer antibiotics penetrate very poorly into the bone in cases of osteomyelitis. These beads are clinically useful in treating difficult cases of osteomyelitis serving the function not only of antimicrobial release but also acting as a spacer preserving bone length. A major factor which limits the use of the beads outside of orthopedic circles is that they require eventual operative removal.

Sustained local concentrations of agents can be achieved by drug incorporation into liposomes. Delgado et al[12] showed that intraperitoneal liposome doxorubicin administered in patients with advanced ovarian cancer results in high concentrations of drug in the peritoneum with very low systemic concentrations. These pharmacologic findings were supported by he clinical decrease in systemic toxicities. Peyman and associates have demonstrated the clinical effectiveness of the prolonged local levels of drugs resulting from liposomal encapsulation. They found that intravitreal application of liposomal antibiotics was successful in treating otherwise refractory chronic intraocular infections.[13] Rutenfranz et al showed that intramuscularly injected liposome encapsulated interferon gamma resulted in both sustained local concentration of interferon but also sustained release of interferon systemically.[14] The sustained release resulting from liposomal encapsulation accounted, at least in part, for the therapeutic advantage seen in our laboratory treatment of experimental peritonitis to be discussed later.[15] With local liposomal delivery, the resultant decrease in absorption can also decrease systemic toxicity as demonstrated in the clinical studies by Delgado and Peyman. Because of their biodegradable nature, the local administration of agents with liposomes also eliminates any need for subsequent removal as is the case with methylmethacrylate beads.

Experimental Studies

To further examine the potential of local liposome antibiotic use in surgical settings we designed a study to answer the questions: 1) does liposomal delivery of antimicrobials result in any therapeutic advantage when compared to locally applied free antimicrobials in a contaminated subcutaneous wound model; 2) does the presence of an infection alter the absorption and local concentration of liposomal antimicrobials; and, 3) does subcutaneous administration of liposomal antibiotics decrease systemic absorption and sustain local concentrations when compared to the unencapsulated form. In addition, the pharmacology of the liposomes and the encapsulated agent was examined separately after subcutaneous administration. The model of subcutaneous infection we chose mirrored a contaminated surgical wound.

During construction of the proprietary liposomes, a radioactive free fatty acid (I^{125} phenyldecanoic acid) was added to radiolabel the liposomes. After formulation, SephadexR G-50 gel column filtration studies confirmed the total incorporation of the radiolabelled free fatty acid into the liposomes. Column filtration was repeated two weeks after the study to confirm the persistence of the radioactive tag. (Fig. 1) The specific radioactivity of the preparation was 34.7 microcurie per milliliter. Tobramycin was encapsulated into the liposomes resulting in a concentration of 16 mg of tobramycin per milliliter.

Adult Sprague Dawley rats were deeply anesthetized with methoxyflurane and each rat had a one centimeter incision made over the mid back. Animals were then either infected with 108 CFU *Pseudomonas aeruginosa* (group 1, N=102) or left uninfected (group 2, N=35). Prior to closure of the wound, the infected animals were treated

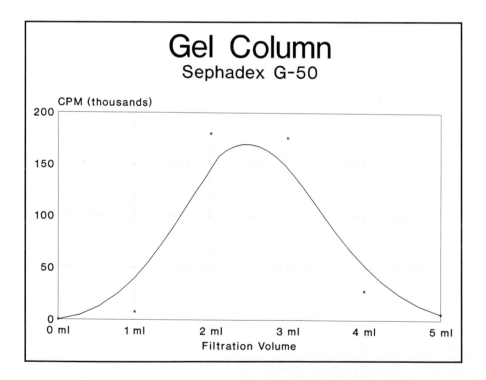

Fig. 1. The gel column filtration of the liposome preparation. There was no residual radiation in the gel column after 5 ml of effluent was collected indicating that all of the I^{125} phenyldecanoic acid was incorporated into the liposomes. The column run at two weeks was identical indicating integrity of the tagged liposomes for the duration of the study period.

with local application of either 0.3 ml saline (untreated group 1, UT, N=30), 0.3 ml (5.5 mg) free tobramycin (FT group 1, N=30) or 0.3 ml (5.5 mg) of liposome-encapsulated tobramycin (LET group 1, N= 42). A group of animals not infected were also treated locally with 0.3 ml (5.5 mg) of liposome encapsulated tobramycin (group 2, N=35). The specific radioactivity of the 0.3 ml of LET was 10.41 μCi.

At 24, 48 and 72 hours postoperatively, a subset of animals in each group was sacrificed. Serum and tissue tobramycin concentrations were measured and quantitative tissue cultures were done. Blood cultures were done on 10 randomly selected animals in each of the UT, FT and LET infected groups 4 hours postinfection. Tissue samples for antibiotic assay were homogenized and acid digested for 24 hours prior to assay. A fluorescence polarization technique was used for the antibiotic assay.

Blood, operative site tissue and all splanchnic organs were harvested for assessment of differential radioactivity using a gamma scintillation counter. Differential measurement of radioactivity allowed liposome distribution to be followed separately from that of the encapsulated tobramycin. Radioactivity in an organ or tissue was reported as a percent of the total dose of radioactivity administered. The operative site tissue (skin and muscle) was equally divided and the tissue used for tobramycin assay, bacterial counts and for assessment of local radioactivity. Tissue bacterial counts, tissue tobramycin levels and tissue radioactivity levels were all reported normalized per gram of tissue.

There were three deaths in the UT group (10%), one death in the FT group (3.3%), and one death in the LET group (2.4%). These differences were not significant but did suggest that there could be a survival advantage with liposome antibiotics. This left a total of 130 animals for study.

Blood culture results indicated a fairly localized infection. There was one positive blood culture in the 25 animals randomly selected from the infected rats, occurring in an animal treated with liposome encapsulated tobramycin (LET). This may have represented a contaminant.

Tissue culture data showed that there were significantly lower numbers of colony forming units (CFU) at 24 hours in both the tobramycin treatment groups (FT, LET) when compared to the untreated group; this difference persisted at 48 and 72 hours postinfection. At 48 hours there were significantly fewer numbers of organisms recovered from the tissues of the LET group compared to the FT group and this difference persisted at 72 hours. At twenty-four hours after inoculation, there was no significant difference between the FT and LET groups with regard to the number of colony forming units (p< 0.05, Table 1).

Serum tobramycin levels were low throughout the study with the majority of the specimens having less than 0.16 ug/ml (Table 2). Statistically there were significantly higher levels of serum tobramycin in both the LET treated rats (infected and noninfected) when compared to the FT group at both 24 and 48 hours but not at 72 hours. The noninfected rats treated with LET had higher levels of serum tobramycin at 24 and 48 hours than the infected animals. At 72 hours both the infected and noninfected liposome treated animals had similar serum tobramycin levels. Tissue tobramycin levels were significantly lower in the FT group than in the LET infected group at 24 (p=0.02), 48 (p=0.03) and 72 (p=0.01) hours after inoculation (Table 2). The infected group treated with LET tended to have higher tissue levels of tobramycin than the noninfected group treated with LET at 24, 48, and 72 hours, but these differences were not statistically significant.

The radioactive liposome distribution data parallelled the tobramycin serum and tissue data. The majority of the radioactivity was recovered from the site of inoculation; an average of 38.4 % of the total administered

Table 1. Colony Forming Units of Pseudomonas Aeruginosa Per Gram of Tissue, Expressed as a Natural Log.

	Untreated		Free Tobramycin		Liposome Tobramycin	
	Ave CFU	SEM	Ave CFU	SEM	Ave CFU	SEM
24 hrs	6.785	0.219	5.505	0.302	4.882	0.579
	(n=9)		(n=10)		(n=12)	
48 hrs	7.148	0.171	3.500	0.213	1.501	0.388
	(n=10)		(n=10)		(n=15)	
72 hrs	7.049	0.472	3.535	0.246	2.176	0.396
	(n=7)		(n=9)		(n=13)	

Table 2. The Serum and Tissue Tobramycin Levels

	Free Tobra		Serum Tobramycin (μgm/ml) Liposome Tobra (infected)		Liposome Tobra (non-infected)	
	Ave	SEM	Ave	SEM	Ave	SEM
24 hrs	0.016	0.008	0.049	0.009	0.082	0.008
	(n=9)		(n=9)		(n=9)	
48 hrs	0	0	0.046	0.009	0.077	0.006
	(n=10)		(n=13)		(n=12)	
72 hrs	0.100	0.008	0.103	0.005	0.089	0.004
	(n=9)		(n=12)		(n=12)	

	Free Tobra		Tissue Tobramycin (μgm/gm) Liposome Tobra (infected)		Liposome Tobra (non-infected)	
	Ave	SEM	Ave	SEM	Ave	SEM
24 hrs	1.89	0.41	4.89	0.84	4.11	0.83
	(n=8)		(n=9)		(n=9)	
48 hrs	0.92	0.19	5.31	1.07	2.92	0.60
	(n=10)		(n=12)		(n=12)	
72 hrs	1.17	0.33	5.60	0.89	4.04	0.70
	(n=6)		(n=12)		(n=12)	

dose was recovered. The majority (>90%) of the recovered radioactivity was in the site of inoculation at each collection time; however, the organ distribution of radiolabelled liposomes was uneven (Table 3). Excluding the tissue at the site of administration, both the LET infected and LET noninfected groups had significantly higher levels of

Table 3. Organ Distribution of Radiolabeled Liposomes

	Infected	Non-Infected
24 hrs	**liver sd:** all others **intest sd:** blood, spleen, skelm	**liver sd:** all others **intest sd:** blood, spleen, skelm heart **kidney sd:** skelms
48 hrs	**liver sd:** all others **intest sd:** blood, spleen, skelm heart, kidney, lung, stomach	**liver sd:** all others
72 hrs	**liver sd:** all others **fat sd:** blood, spleen, skelm heart, stomach	**liver sd:** all others **fat sd:** blood, spleen, skelm heart, stomach, lung stomach

sd - significantly different

radioactive liposomes recovered from the liver than the other tissues at 24, 48 and 72 hours. There were no significant differences in organ distribution of radiolabeled liposomes between the LET infected and noninfected groups. (Fig. 2)

In this study the local application of tobramycin resulted in a decrease in the number of CFU compared with the CFU measured in the untreated infected animals; furthermore, liposome encapsulated tobramycin decreased the number of colony forming units to an even greater extent than free tobramycin. One obvious explanation for the enhanced efficacy observed with LET is the sustained local antibiotic concentration compared with that observed after administration of free antibiotic.

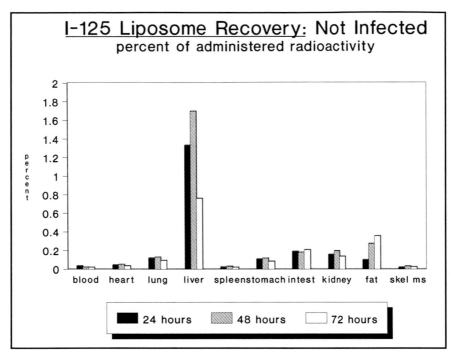

Fig. 2. Graphic representation of the tissue and organ distribution of radiolabeled liposomes expressed as the percent of the administered dose of radiation.

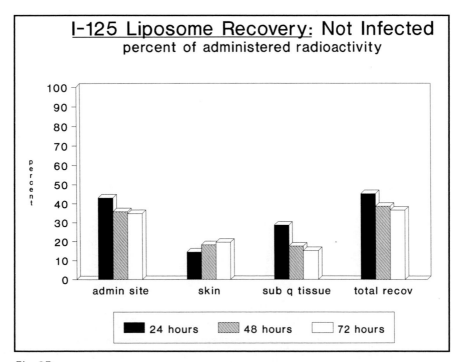

Fig. 2B

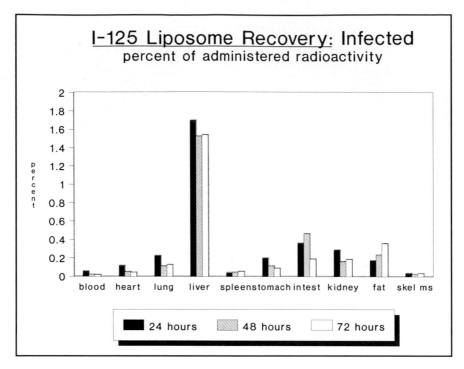

Fig. 2C. Refer to page 31, Fig. 2A.

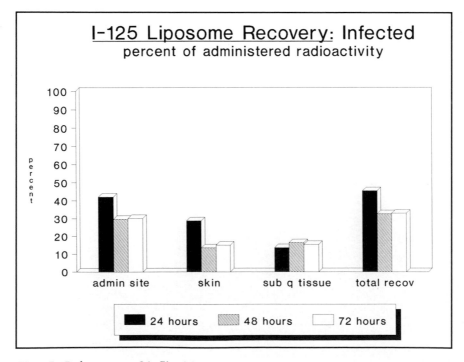

Fig. 2D. Refer to page 31, Fig. 2A.

This sustained concentration would result in antimicrobial being in contact with infecting organism for a longer time interval. While serum antibiotic levels were low in both the FT and LET groups there was a statistically higher serum level in the LET group. The higher tissue tobramycin levels would indicate the liposome tobramycin remained at the site of inoculation in higher concentrations than the free drug. This would necessarily result in more sustained release of antibiotic into the serum in the LET groups accounting for the higher serum levels of drug.

It would be expected that the animals treated with free tobramycin had high serum drug concentrations initially as the tobramycin was rapidly absorbed from the site; it is likely that the timing of our serum tobramycin measurements missed this peak. An attempt was made to compensate for this error by measuring urinary excretion of tobramycin in the first 24 hours posttreatment; however, technical problems precluded accurate assessment of these data.

RES Liposomal Uptake

The antimicrobial advantage of liposome entrapped tobramycin secondary to high local concentration of drug was further confirmed by the differential organ distribution of radiolabeled liposomes. The greatest concentrations of labeled liposomes was at the administration site; in addition, there was a large concentration of labeled liposomes in the liver, which is the largest reticuloendothelial (RES) organ in the rat. Liposomes are known to be recognized and sequestered by the RES.[16,17] Segal et al showed that intravenously applied liposomes were sequestered in reticuloendothelial organs.[18] Hirano and Hunt showed the reticuloendothelial destination of liposomes which were given intraperitoneally.[19] Khato et al and Jackson both showed that local intramuscular and subcutaneous application of liposomes resulted regional lymphatic uptake.[20,21] With local administration RES uptake of liposomal agent would result, in our study, in the concentration of LET in the same sites as the infecting organisms which are also cleared via RES pathways. Alving et al demonstrated the positive influence of this property in the treatment of leishmaniasis with liposome encapsulated drugs.[22]

At the cellular level, liposome can be engulfed by and fuse with phagocytic cells producing an active intracellular delivery of the encapsulated antimicrobial. The resulting intracellular concentration of antibiotic can be higher than that which would be possible on the basis of a concentration gradient diffusion alone.[23-27] Organisms engulfed by phagocytic cells but not killed could, therefore, be exposed to higher concentrations of liposome entrapped drug than free drug. This is the rationale for the use of liposome drugs to treat intracellular pathogens. The extent to which this property has an effect on the treatment of more conventional infections, like that of our model, is unclear.

DISTRIBUTION IN SUBCUTANEOUS FAT

In addition to the higher concentrations of radiolabeled liposomes, in our study, at the administration site and in the liver, there were also high concentrations documented in the peripheral fat. This likely represents the breakdown of the liposomes into constituent fatty acids and the incorporation of these radiolabeled fatty acids into the peripheral fat pool. The total peripheral fat pool incorporation can be extrapolated by calculating the total body fat in the experimental animals. This is important in our experimental results because it has the effect of increasing the percent recovered radioactivity. In clinical situations this accumulation of liposome components in fat indicates the desirability of using safe, naturally occurring constituents in liposome construction.

Further examination of the data suggests a difference in the pharmacology of the LET in the infected and noninfected rats. This is expected since inflammatory host responses change the capillary permeability and local blood flow. These changes would tend to allow for more rapid clearance of even liposomal agents from an infected site. Countering this is the myriad of inflammatory and phagocytic cells brought to the infected site as a part of the host response. These phagocytic cells would tend to engulf the liposome agent sequestering it in the wound. In addition, the local tissue edema brought on by the milieu of inflammatory mediators tends to inhibit clearance via lymphatic channels further slowing clearance of even the engulfed liposome agent. A decrease in the clearance of LET from the infected site was suggested by the trend for higher tissue levels of antibiotic in the infected rats treated with LET. Decreased clearance in infected rats was further suggested by the increased concentration of labeled liposomes recovered from the inoculation site, supporting the hypothesis that presence of infection and an inflammatory response limits clearance of LET from the site of administration. Interpretation of the data in this way suggests a direct correlation between degree of inflammation and retention of locally applied liposomal entrapped drug. If this correlation is valid, one could predict that with increasing amounts of inflammation, there would be a concomitant increase in the efficacy of liposomal agents.

Prophylaxis and Treatment

The use of prophylactic antibiotics in certain surgical procedures is a well documented way to decrease infectious complications. The timing of systemic antibiotic administration is of the utmost importance in prophylaxis in order to obtain adequate tissue levels of antibiotic prior to contamination. While local antibiotic administration is used for the prophylaxis of sites where contamination is suspected, there is conflicting data concerning routine local antibiotic use.[28,29,2,15,7,8,30-32] One reason for the variable results with local antibiotic administration is that locally administered antibiotics are rapidly absorbed from the site of administration, limiting both prophylactic and treatment effectiveness. Both continuous irrigation[33] and incorporation into methylmethacrylate beads[11] have been used to overcome rapid absorption of locally applied antibiotics and to maintain high local drug concentrations. However, neither approach has proven applicable to routine operative prophylaxis. A biodegradable delivery system, such as liposomes, achieves sustained local antimicrobial concentrations and is applicable in most surgical wounds.

Liposome delivery of local soft tissue antibiotics may find a clinical use in surgical wound prophylaxis. Liposome delivery of tobramycin in our model of surgical wound contamination and infection reduced the number of organisms to a greater extent than the unencapsulated tobramycin. In animals treated with LET, the number of CFU was consistently less than the 1×10^5 organisms per gram of tissue that is considered critical for invasive infection. The potential for liposome agent use in surgical prophylaxis is huge since antimicrobials used in a prophylactic manner often make up the majority of antimicrobial use in a busy surgical hospital. Advantages of liposomal encapsulation in surgical wound prophylaxis include the achievement of adequate local levels of antimicrobials reliably at the time of surgery and increased efficacy of the antibiotic. As opposed to systemic antibiotic prophylaxis, which is less effective when given after the start of a procedure, our animal data suggests that local liposomal antibiotics may be given after contamination has occurred and still be effective. By allowing for more specific selection of cases in which prophylaxis is warranted, liposome antibiotics could potentially lessen the total cost as well as the total amount of antimicrobials used. Currently, antibiotics are widely over used because prophylactic antibiotics must be given prior to the start of surgery to achieve maximal benefit and it is often impossible to predict preoperatively whether a procedure will result in a contaminated wound. The local delivery of prophylactic agents which otherwise have a high risk of systemic side effects may be safer when applied locally within liposomes. One such agent is the commonly used triple antibiotic irrigation solution which can potentiate neuromuscular blocking agents. Liposome delivery would tend to limit this side effect by limiting absorption.

In other clinical settings local liposome antibiotics may also offer some advantages. For instance, an infected vascular conduit or orthopedic implant is often associated with catastrophic results. Systemic antibiotics generally fail to clear these infections, and frequently the prosthesis must be removed to eradicate the infection. Also, at times, poorlyvascularized or devitalized tissue precludes adequate penetration of systemically applied antibiotics as in the examples of necrotizing pancreatitis and massive soft tissue trauma. While several technics are currently used to achieve sustained local concentrations of antibiotics, each technic has severe limitations. In the treatment of osteomyelitis, methylmethacrylate beads are used to achieve high local concentrations of antibiotics with good results. However, these beads must be removed and are not applicable in most routine surgical wounds. Continuous antibiotic irrigation through indwelling catheters can also be effective in some difficult to eradicate infections but is unwieldy for routine use and can result in systemic toxicities. However, both of these technics result in a high local concentration of antibiotic at the site of the infected prosthesis or poorly vascularized tissue and therefore may prevent the need to remove an infected prosthesis and salvage situations otherwise recalcitrant to therapy. Local liposome antibiotic therapy also has the potential to resolve these difficult clinical situations. Our data confirm that the liposome delivery system achieves the goal of sustained local drug concentrations. The application of this method of drug delivery for antimicrobial prophylaxis and to the treatment of some of these more difficult infectious situations warrants further study.

REFERENCES

1. Polk H, Trachtenberg L, Finn MP. Antibiotic activity in surgical incisions: The basis for prophylaxis in selected operations. JAMA 1980; 244:1353-1354.
2. Finch DRA, Taylor L, Morris PJ. Wound sepsis following gastrointestinal surgery: A comparison of topical and two dose systemic cephradine. Br J Surg 1979; 66:580-582.
3. Burke JF. The effective period of preventative antibiotic action in experimental incisions and dermal lesions. Surgery 1960; 50:161-168.
4. Bergamini TM, Polk HC. Pharmacodynamics of antibiotic penetration of tissue and surgical prophylaxis. Surg Gynec Obstet 1989; 168:283-289.
5. Pitt HA, Postier RG, Gadacz TR, Cameron JL. The role of topical antibiotics in 'high-risk' biliary surgery. Surgery 1982; 91:518-524.
6. Pitt HA, Postier RG, MacGowan WAL et al. Prophylactic antibiotics in vascular surgery: Topical, systemic, or both. Ann Surg 1980; 192(3):356-34.
7. Sarr MG, Parikh KJ, Sanfey H, Minkin SL, Cameron JL. Topical antibiotics in the high-risk biliary patient: A prospective, randomized study. Amer J Surg 1988; 155:337-341.
8. Uander Salm TJ, Okoke ON et al. Reduction of sternal wound infection by application of topical vancomycin. J Thorac Cardiovasc Surg 1989; 98:618-22.
9. ReMine SG, Organ CH. Local Antibiotic infusion after primary closure of heavily contaminated wounds. Infect Surg 1989; 8,2:55-59
10. Florey ME, Williams REO. Hand infections treated with penicillin. 1944; 1:73-81.
11. Wilson KJ, Cierny G, Adams K, Mader JT. Comparative evaluation of the diffusion of tobramycin and cefotaxime out of antibiotic impregnated polymethylmethacrylate beads. J Orthop Res 1988; 6,2:279-286.
12. Delgado G et al. A phase I/II study of intraperitoneally administered doxorubicin entrapped in cardiolipin liposomes in patients with ovarian cancer. Am J Obstet Gynec 1989; 160(4):812-819.
13. Peyman GA, Charles HC, Liu KR, Khoobehi B, Niesman M. Intravitreal liposome-encapsulated drugs: A preliminary human report. Int Ophth 1988; 12:175-182.
14. Rutenfranz I, Bauer A, Kirchner H. Pharmacokinetic study of liposome-encapsulated human interferon-gamma after intravenous and intramuscular injection in mice. J Interferon Res 1990; 10:337-341.
15. Price CI, Horton JW, Baxter CR. Enhanced effectiveness of intraperitoneal antibiotics administered via liposomal carrier. Arch Surg 1989; 124:1411-1415.
16. Ostro MJ. Liposomes. Sci Amer 1987; 256:102-111.
17. Juliano RL. Liposomes as drug carriers in the therapy of infectious diseases. Horiz Biochem Biophys. 1989; 9:249-279.
18. Segal AW, Willis EJ, Richmond JE, Slavin G, Black CD, Gregoriadis G. Morphological observations on the cellular and subcellular destination of intravenously administered liposomes. Br J Exp Pathol 1974; 55:320-327.
19. Hirano K, Hunt CA. Lymphatic transport of liposome-encapsulated agents:effects of liposome size following intraperitoneal administration. J Pharm Sci 1985; 17:915-921.
20. Khato J, de Campo A, Sieber S. Cancer activity of sonicated small liposomes containing melphan to regional lymph nodes of rats. Pharmacology 1983; 6:260-270.

21. Jackson AJ. Intramuscular absorption and regional lymphatic uptake of liposome-entrapped inulin. Drug Metab Disp 1981; 9(6):535-540.

22. Alving CR, Steck EA, Chapman WL, Waits VB, Hendricks LD, Swartz GM, Hanson WL. Therapy of Leishmaniasis: Superior efficacies of liposome-encapsulated drugs. Proc Natl Acad Sci 1978; 75(6):2959-2963.

23. Fountain MW, Dees C, Schulte RP. Intracellular killing of S aureus by liposomes containing aminoglycosides: A comparison of liposome surface charge and liposome entrapment of amikacin, tobramycin, gentamicin and kanamycin. Curr Microbiol 1981; 6:373-376.

24. Nicoletti P, Lenk RP, Popescu MC, Swenson CE. Efficacy of various treatment regimes, using liposomal streptomycin in cows with brucellosis. Am J Vet Res 1989; 50(7):1004-1007.

25. Coune A. Liposomes as drug delivery system in the treatment of infectious diseases: Potential applications and clinical experience. Infection 1988; 16:141-147.

26. MacLeod DL, Prescott JF. The use of liposomaly-entrapped gentamicin in the treatment of bovine staphylococcus aureus mastitis. Can J Vet Res 1988; 52:445-450.

27. Bakker-Woudenberg IAJM, Lokerse AF, Roerdink FH, Regts D, Michel MF. Free versus liposome-entrapped ampicillin in treatment of infection due to Listeria monocytogenes in normal and athymic (nude) mice. J Infec Dis 1985; 151(5):917-924.

28. Lord JW, LaRaja RD, Daliana M, Gordon MT. Prophylactic antibiotic irrigation in gastric, biliary, and colon surgery. Am J Surg 1983; 145:209-212.

29. Raahave D, Hesselfeldt P, Pederson T, Zachariassen A, Kann D, Hansen OH. No effect of topical ampicillin prophylaxis in elective operations of the colon or rectum. Surg Gynecol Obstet 1989; 168:112-114.

30. Evans C, Pollock AV, Rosenberg IL. The reduction of surgical wound infections by topical cephaloridine: A controlled clinical trial. Br J Surg 1974; 61:133-135.

31. Halasz NA. Wound infection and topical antibiotics: The surgeons dilemma. Arch Surg 1977; 112:1240-1244.

32. Hares MM, Hegarty MA, Warlow J et al. A controlled trial to compare systemic and intra-incisional cefuroxime prophylaxis in high risk gastric surgery. Br J Surg 1981; 68:276-280.

33. Liebhoff AR, Soroff HS. The treatment of generalized peritonitis by closed peritoneal lavage: A critical review of the literature. Arch Surg 1987; 122:1005-1010.

CHAPTER 5

LIPOSOME ANTIBIOTICS IN INTRAPERITONEAL INFECTIONS

For many years, local installation of antibiotics into the peritoneal cavity has been used for infection prophylaxis as well as for treatment despite the lack of conclusive data to support local antibiotic use.[1-7] The detractors of peritoneal antibiotic use cite the rapid absorption from the large surface area of the peritoneal cavity which rapidly decreases the concentration of local antibiotic and obviates any advantages over systemic administration of drug.[8] In addition, this rapid absorption can result in unintended side effects or toxicities. Stone emphasized that exposure of the large surface area of the peritoneum to a bacterial challenge is similar to that observed with the large loss of skin and the loss of the protective barrier after burn injury.[9] The treatment of burns with parenteral antibiotics has uniformly failed to either eradicate pathogens or to prevent bacterial colonization of burn wounds. Local application of appropriate antimicrobials, however, has prevented bacterial colonization, has reduced the subsequent development of sepsis, and has impacted positively on survival.[9] Since the pathophysiology of peritonitis is similar to that of large burns the use of an adjunct to delay antibiotic absorption and provide for a longer local antimicrobial effect in the peritoneal cavity has sound clinical precedent.

Hirano and Hunt experimentally showed liposome bound sucrose to persist in the peritoneal cavity, resisting systemic absorption. Additionally, the clearance of liposome bound sucrose from the peritoneal cavity occurred through the lymphatic system rather than by direct absorption into the bloodstream.[10] Delgado et al and Sells et al also demonstrated decreased absorption of liposome agent from the peritoneal cavity.[11,12] With this background, the effects of a water soluble antibiotic was examined experimentally in a model of peritonitis.

Adult Sprague-Dawley rats had peritonitis initiated using the method described by Nichols et al.[13] This consisted of a 0.3 ml fecal slurry/barium sulfate inoculum of a known microbial composition that was placed in a gelatin capsule and inserted into the abdominal cavity through a lower midline incision. The wounds were closed in two layers with nonabsorbable sutures in the fascia and surgical staples in the skin.

One milliliter of blood was drawn from each animal 4 and 24 hours after inoculum by cardiac puncture under sterile conditions; this sample was used for blood cultures and WBC counts. Quantitative cultures were done using McConkey's agar; WBC counts were determined with a coulter counter.

After induction of peritonitis, animals were divided into four groups: group 1 (N=20) untreated; group 2 (N=15) treated with cefoxitin 1.5 mg/kg intramuscularly at the time of peritoneal contamination; group 3 (N=15) treated with 1.5 mg/kg of cefoxitin given intraperitoneally (IP) at the time of contamination and group 4 was treated with 1.5 mg/kg of liposome entrapped cefoxitin given intraperitoneally at the time of peritoneal contamination. Additional rats (N=17) received either free or liposome entrapped tobramycin given IP at the time of peritoneal contamination and the had serial serum tobramycin levels determined using an immunoassay technic.

At seven days, all surviving rats in groups 1-4 were killed and the intraperitoneal abscess number determined. Only the abscesses that were discrete and greater in size than 1 mm were counted.

The blood cultures for all untreated animals were positive and *Escherichia coli* was the predominant organism isolated. Animals treated with either IM or IP free cefoxitin had significantly fewer numbers of organisms in the blood than the untreated animals 4 and 24 hours after contamination. However, neither IM nor IP routes of free cefoxitin administration demonstrated an advantage in terms of lower bacterial counts when compared with one another. Intraperitoneal administration of liposome-encapsulated cefoxitin resulted in a significant reduction in bacteremia at both 4 and 24 hours compared with untreated animals as well as animals treated with either IM or IP free cefoxitin. (Table 1)

None of the blood cultures for the untreated animals were negative. Animals treated with IM cefoxitin had 16% negative cultures compared with 6% negative cultures in animals treated with IP cefoxitin. Cultures were negative in 56% of the animals with IP liposomes.(Fig. 1) To more accurately assess the level of bacteremia and for statistical considerations, the blood culture results were recalculated and the cultures with negative results were excluded. The relative number of organisms in each group remained consistent and confirmed fewer circulating organisms after all treatment regimens compared with the untreated groups. Furthermore, liposome-encapsulated IP cefoxitin was associated with fewer organisms than either treatment group at both 4 and 24 hours after inoculation.

Table 1. Blood Culture Results Expressed as the Natural Log of the Number of Colony Forming Units per Milliliter of Serum

	4 hrs	24 hrs
Untreated	3.90±0.08 (n=15)	5.69±0.46 (n=7)
I.M.Cefoxitin (1.5 mg/kg)	2.80±0.34* (n=14)	3.38±0.80* (n=8)
I.P.Cefoxitin (1.5mg/kg)	2.47±0.16* (n=10)	2.39±0.60* (n=10)
I.P.Liposome Cefoxitin (1.5 mg/kg)	0.48±0.20* (n=10)	1.39±0.53* (n=10)

*Significantly different from untreated, $p \leq 0.05$; ANOVA

Survival data paralleled the blood culture results. Untreated animals had a 90% mortality. There was a 33% survival rate in animals treated with IM cefoxitin compared with 80% survival in animals treated with IP cefoxitin. Treatment with liposome-encapsulated cefoxitin produced 100% survival. Most deaths occurred in the first 48 hours after inoculation, and no deaths occurred after 4 days. (Fig. 2)

The mean (± SEM) number of abscesses for animals treated with IM cefoxitin was 9.2 ± 2.32. Animals treated with IP cefoxitin had 10.33 ± 1.68 abscesses while those treated with IP liposome cefoxitin had 6.86 ± 0.79 abscesses. The difference in treatment groups was not statistically significant. Assessment of untreated animals was not possible because too few animals survived a sufficient time for abscess formation. Examination of the peripheral WBC counts showed that untreated animals had statistically lower numbers of circulating WBC at 4 hours compared with treatment groups. At 24 hours this difference was ablated, and WBC counts were similar in all groups.

In animals given free IP tobramycin (N = 7), the mean (± SEM) serum tobramycin level 4 hours after fecal inoculation was 4.3 ± 0.83 mg/L; at 24 hours after inoculation, serum tobramycin concentration level was 1.1 ± 0.35 mg/L. Animals given IP liposome-encapsulated tobramycin (N = 10) had a 4 hour serum level of 1.15 ± 0.27 mg/L and 24 hour levels of 2.02 ± 028 mg/L. (Fig. 3)

Additional animals were treated with the empty liposome preparation to see if the liposomes themselves were in some way impacting on the results. The animals treated with the empty liposomes had the same mortality and blood culture results as did the untreated animals.

Local antibiotic use, especially in peritoneal cavity irrigation, is a time-honored tradition of many surgeons. Several studies have supported this surgical practice by demonstrating a decreased incidence of infectious complications with local antibiotics use.[1-5,14,6] Other surgeons have been much less enthusiastic regarding local antibiotic use in the peritoneum. Objections to local antibiotic use include rapid absorption from a large surface area such as the peritoneal cavity, a rapid decrease in local antibiotic concentration, as well as systemic side effects.

As discussed earlier in this monograph, liposomes have been used to deliver numerous pharmacologic agents in various experimental and clinical settings. Both experimental and clinical trials have explored the use of antineoplastic and antimicrobial agents delivered by liposomal carriers, and several therapeutic advantages have been reported.[15,16,17,12,18]

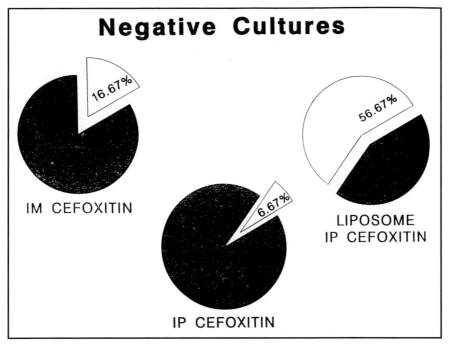

Fig. 1. Percent of blood cultures negative at both 4 and 24 hours. There were no negative cultures in the untreated animals.

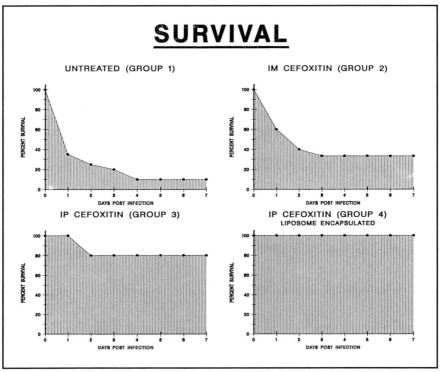

Fig. 2. Percent of animals surviving to seven days. There were no deaths in the liposome treated animals.

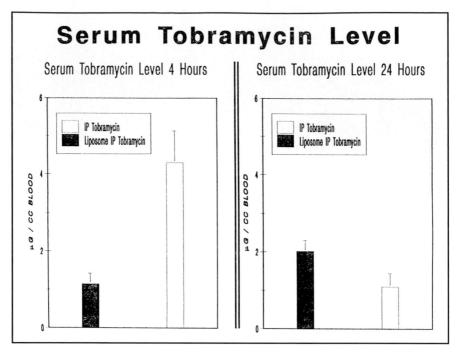

Fig. 3. The tobramycin levels in the serum after intraperitoneal administration showing more systemic absorption when unencapsulated drug is given.

In addition, immunomodulaters and vaccines have also been delivered via liposomal carriers, and the results have been encouraging.[15] The efficacy of liposomes in these studies results from either increasing the concentration of a pharmacologic agent in an organ system (for example, the reticuloendothelial system or lung) or preventing the pharmacologic agent from concentrating in a organ where toxicity is a concern.[12]

One aspect of liposome pharmacodynamics that has not been fully exploited is the ability of liposomes to decrease systemic absorption of agents from a body cavity where persistence of a drug would be highly advantageous. Hirano and Hunt[10] implied this advantage in their demonstration of lymphatic clearance of liposome-bound sucrose from the peritoneal cavity and the persistence of sucrose in the peritoneum. Consideration of these data suggested to us the possible advantages of local delivery of antibiotics in intraabdominal infections and contamination. In our peritonitis study, liposomal delivery of cefoxitin was associated with a greater number of negative blood cultures, lower levels of bacteremia, and increased survival compared with untreated animals or compared with animals treated with either free intramuscular cefoxitin or free intraperitoneal cefoxitin, indicating greater persistence of cefoxitin in the peritoneal cavity.

The tobramycin data support peritoneal persistence of intraperitoneally administered liposome encapsulated antibiotics and may explain the increased efficacy of cefoxitin when delivered by liposomes. Intraperitoneal delivery of free tobramycin resulted in an initial increase in serum tobramycin levels 4 hours after administration followed by a marked reduction by 24 hours. In contrast, liposome tobramycin produced significantly higher serum tobramycin levels at 4 hours and this effect was sustained 24 hours after inoculation. These data indicate the persistence of drugs in the peritoneal cavity and confirm slow release with liposomal delivery of antibiotics. The trend toward fewer abscesses in animals treated with IP liposome cefoxitin further support the peritoneal persistence of antibiotics after liposome delivery.

The increased efficacy seen with liposome cefoxitin may also result from the recognition and uptake of liposomes by phagocytic cells in the peritoneum and subsequent sequestration in the reticuloendothelial system. Since these same phagocytic cells and lymphatic components would also be responsible for bacterial clearance, the liposomal delivery of antimicrobials in this situation concentrates antibiotics at the site of organism concentration. Finally, while the exact interaction of liposomes with the immune system is not yet defined, some reports have indicated immune enhancement with liposome administration.[15] The lack of quantitative difference in WBC count between animals given free cefoxitin and liposome-encapsulated cefoxitin suggests that liposomes have no effect on this gross measure of host response. Moreover empty liposomes had no effect on host survival.

In our model of peritonitis, IP administration of liposome-encapsulated cefoxitin resulted, after a single dose, in decreased levels of bacteremia and an increased number of negative cultures compared with conventional IM and IP cefoxitin administration. In addition, IP liposomal cefoxitin significantly improved survival.

Clinically, delivery of liposome encapsulated antibiotic may offer advantages in situations such as penetrating trauma, where contamination is a major concern and adequate levels of antibiotics cannot be obtained prior to contamination as would be the case with elective surgical procedure. Liposome encapsulated antibiotics may also be beneficial in the treatment of abscesses and conditions such as necrotizing pancreatitis. In these situations systemic antibiotics often fail to achieve adequate concentrations in the effected tissue. Combining drainage and debridement with locally increased antibiotic concentration after liposomal delivery may have a beneficial effect compared with systemic antibiotics alone. Moreover, with the ever increasing use of minimally invasive surgical technics and radiologic drainage of abscesses, wide surgical debridement may become less common. The use of one of these less invasive procedures along with liposome antibiotics may allow for the successful treatment of situations which may otherwise require more aggressive surgery. Also, peritoneal dialysis patients, who commonly present with peritonitis, are currently treated with a combination of intraperitoneal and systemic antibiotics. They may be more appropriately treated with intraperitoneal liposomal antibiotics. In these patients, liposome antibiotics could not only be more effective at treating the peritonitis but also in sterilizing the peritoneal dialysis catheter. These catheters often harbor glycocalyx forming organisms which provide a source for recurrent episodes of infection. While extrapolation of any animal data to clinical situations must be approached with caution, liposomal delivery of local antibiotics appears to offer real advantages.

REFERENCES

1. Lord JW, LaRaja RD, Daliana M, Gordon MT. Prophylactic antibiotic irrigation in gastric, biliary, and colon surgery. Am J Surg 1983; 145:209-212.
2. Bergamini TM, Polk HC. Pharmacodynamics of antibiotic penetration of tissue and surgical prophylaxis. Surg Gynecol Obstet 1989; 168:283-289.
3. Pitt HA, Postier RG, Gadacz TR, Cameron JL. The role of topical antibiotics in 'high-risk' biliary surgery. Surgery 1982; 91:518-524.
4. Sarr MG. Parikh KJ. Sanfey H, Minkin SL, Cameron JL. Topical antibiotics in the high-risk biliary patient: A prospective, randomized study. Amer J Surg 1988; 155:337-341.
5. Evans C, Pollock AV, Rosenberg IL. The reduction of surgical wound infections by topical cephaloridine: A controlled clinical trial. Br J Surg 1974; 61:133-135.
6. Brockenbrough EC, Moylan JA. Treatment of contaminated surgical wounds with a topical antibiotic: A double-blind study of 240 patients. Am Surg 1969; 35:789-792.

7. Diagnosis and management of peritonitis in continuous ambulatory peritoneal dialysis: Report of a working party of the british society for antimicrobial chemotherapy. Lancet 1987; i:845-849.

8. Morse GD, Apicella MA, Walshe JJ. Absorption of intraperitoneal antibiotics. Drug Intell Clin Pharm 1988; 22:58-61.

9. Stone HH. Invited commentary. Arch Surg 1989; 124:1415

10. Hirano K, Hunt CA. Lymhatic transport of liposome-encapsulated agents: Effects of liposome size following intraperitoneal administration. J Pharm Sci.1985; 17:915-921.

11. Delgado G et al. A phase I/II study of intraperitoneally administered doxorubicin entrapped in cardiolipin liposomes in patients with ovarian cancer. Amer J Obstet Gynecol 1989; 160(4):812-819.

12. Sells RA, Gilmore IT, Owen RR, New RRC, Stringer RE. Reduction in doxorubicin toxicity following liposomal delivery. Cancer Treat Rev 1987; 14:383-387.

13. Nichols RL, Smith JW, Balthazar ER. Peritonitis and intraabdominal abscess: An experimental model for the evaluation of human disease. J Surg Res 1978; 25:129-134.

14. Halasz NA. Wound infection and topical antibiotics: The surgeons dilemma. Arch Surg 1977; 112:1240-1244.

15. Ostro MJ.Liposomes. Sci Amer 1987; 256:102-111.

16. Fountain MW, Weiss SJ, Fountain AG Shen A, Lenk RP. Treatment of brucella canis and brucella abortus in vitro and in vivo by stable pleurilamellar vesicles encapsulated aminoglycoside. J Infect Dis 1985; 152:529-536.

17. Amber L. Liposome encapsulation enhances antibiotic activity. Infec Dis Today 1988; 2:26.

18. Lopez-Berestein G, Fainstein V, Hopfer R et al. Liposomal amphotericin B for the treatment of systemic fungal infections in patients with cancer: A preliminary study. J Infect Dis 1985;151:704-710.

CHAPTER 6

LIPOSOMES IN BURN THERAPY

Topical antimicrobial therapy is an integral part of burn wound management. A number of topical agents are used to decrease burn wound bacterial counts including silver nitrate, nitrofurantoin, aminoglycosides, mafenide and silver sulfadiazine (SSD). While these agents are effective, the short duration of activity as well as adverse effects from systemic absorption may limit efficacy in the clinical setting. Also frequent application of these creams and ointments necessitate cleaning to remove carrier residue, resulting in increased patient discomfort and inefficient resource utilization.

The prolonged local drug concentrations and enhanced effectiveness resulting from liposomal encapsulation and demonstrated by many studies are favorable attributes for topical agents used in the treatment of burn wounds.[1-6] Liposomal delivery of antimicrobial agents to burn wounds may reduce adverse reactions by decreasing the systemic absorption of an agent; furthermore, the biodegradable nature of liposomes may reduce the need to actively remove drug carrier residue as is required with currently used agents.

We undertook a study to determine if topical liposomal delivery of antimicrobials to burn wounds resulted in prolonged tissue drug concentrations. The efficacy of therapy was not examined because of previous data confirming at least equal effect of liposomal agents in comparison to conventional ones. The in vivo disposition of the liposomal carrier was examined as well as the pharmacology of the encapsulated antimicrobial when applied topically to a burn wound.

Liposomes were labelled with a radioactive free fatty acid (I^{125} phenyldecanoic acid); the specific radioactivity of the preparation was 34.7 microcurie per milliliter. Tobramycin was encapsulated into the liposomes resulting in a concentration of 16 mg of tobramycin per ml.

Adult Sprague Dawley rats were deeply anesthetized with inhaled methoxyflurane and subjected to a 10% total body surface area full thickness scald burn using the Walker burn model.[7] All animals had 0.3 cc of radiolabeled liposome-encapsulated tobramycin (10.41 microcurie, 5.5 mg tobramycin) applied topically. The wounds were covered with Opsite® dressing. At 24, 48 and 72 hours postburn, a subset of animals was sacrificed, and at these time intervals, serum and tissue tobramycin concentrations were measured. Tissue samples were homogenized and acid digested for 24 hours prior to assay; a fluorescence polarization technique was used for the drug assay. Blood, burn tissue and splanchnic organs were also harvested for assessment of radioactivity using a gamma scintillation counter. Differential measurement of radioactivity allowed liposome distribution to be followed separately from that of the encapsulated tobramycin. Radioactivity in an organ or tissue was reported as a percent of the total administered radioactivity. The burn tissue was equally divided with half used for the tobramycin assay and half for the assessment of radioactivity.

Serum tobramycin concentrations were low at each sampling time.(Fig. 1) No sample concentration exceeded the minimal sensitivity of the assay which was 0.16 μg/ml. Tobramycin concentrations in burn tissue were 113.85+21.49 mcg/gm of tissue during the 72 hour study period.(Fig. 2) The mean tissue concentration 24 hours postburn was 179.25+28.69 μg/gm, while at 48 and 72 hours the drug concentrations were 113.85±21.49 μg/gm and 114.96±25.99 μg/gm, respectively. There were no statistical differences in the drug tissue concentrations after burn injury.

Approximately 66% of the total administered radioactivity was recovered; that greater than 90% of the recovered radioactivity remained either in the burn tissue or on the burn dressing, and peripheral organs and tissues had less than 1.2% of the administered radioactivity. Of the peripheral organs and tissues studied, the liver had significantly

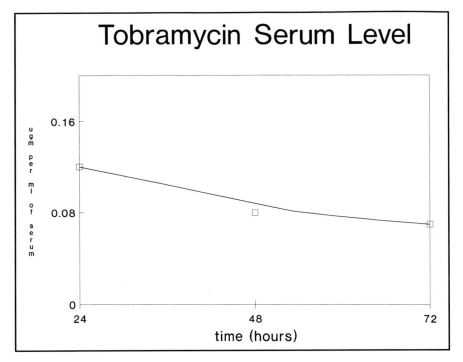

Fig. 1.   *Tobramycin serum levels expressed as µgm/ml.*

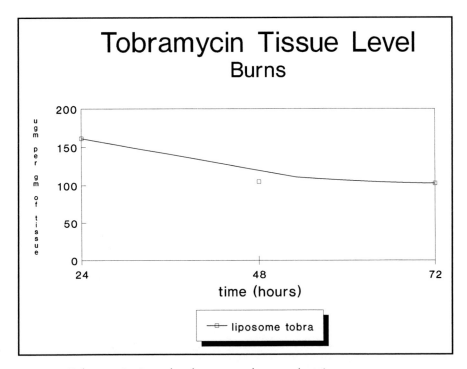

Fig. 2.   *Tobramycin tissue levels expressed as µgm/gm tissue.*

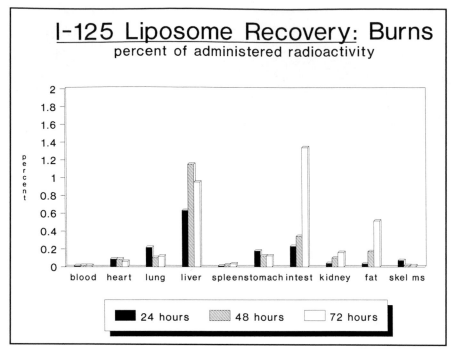

Fig. 3A. Differential radioactivity expressed as a percent of the administered dose of radiation.

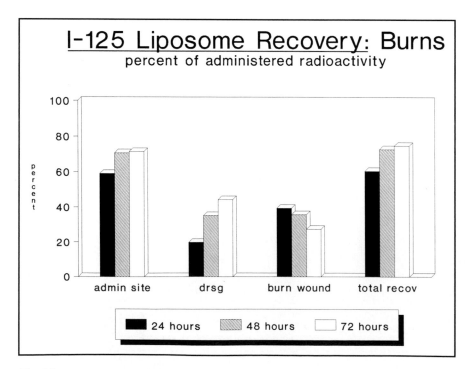

Fig. 3B.

higher levels of radioactivity than most other organs and tissues at all time periods studied (24, 48 and 72 hours, p<0.01). At the 72-hour interval, radioactivity in the peripheral adipose tissue and in the small intestine were similar to that measured in the liver.(Fig. 3)

In peripheral adipose tissue, there was a significantly higher level of radioactivity at 72 hours compared to that noted at 24 and 48 hours (p<0.01). While radioactivity recovered from the kidney and the spleen at 72 hours were significantly greater than that measured at 24 hours (p≤0.01).

Topical antimicrobial therapy is an important aspect of burn wound management. Inadequate local therapy can result in increased septic complications and death. While currently used agents are generally effective, adverse effects from systemic absorption may limit their use in this clinical setting. In addition frequent applications are required to maintain effective antimicrobial action. A drug delivery system that facilitates local persistence of antimicrobial agents on burn wounds without systemic adverse effects has obvious advantages.

There is much data which demonstrate properties of liposomes which would be beneficial in the treatment of burn wounds. For example, our earlier data showed that a single topical dose of liposomal entrapped antimicrobials was equally as effective as multiple topical dosages of free drug.[8] This could allow for less frequent dressing changes in burn patients. Other data from our lab demonstrated that in experimental polymicrobial peritonitis, a single intraperitoneal dose of liposomal entrapped antibiotic was more effective than intraperitoneally applied free drug.[4] Peritonitis has been likened to a large surface area burn. A similar efficacy advantage with liposome antimicrobials applied to burns would be beneficial since large burns still carry a significant mortality from infectious complications. Intraperitoneally applied liposome entrapped doxorubicin has been shown by Delgado and associates to result in higher intraperitoneal concentration of agent with markedly decreased systemic and local toxicity when compared to free intraperitoneal doxirubicin.[2] Like the use of doxorubicin in the peritoneal cavity the decreased absorption of liposomal agents from the burn wound could allow for the use of otherwise toxic dosages or drugs. Peyman et al showed that intravitreal injection of liposome entrapped anti-infective agents had therapeutic advantages in the treatment of various intraocular infections when compared to unencapsulated agents and required less frequent dosing.[9] This again supports the notion that liposome encapsulation would result in less frequent dosing and therefore dressing changes in burned patients. Also, data from our lab showed that a single dose of locally applied liposome entrapped tobramycin decreased tissue bacterial counts to a greater extent than free tobramycin in the treatment of a subcutaneous infection.[10] Surely the enhanced efficacy of liposomal agents seen elsewhere will hold up in burn care which could ultimately result in the salvage of clinical situations that are otherwise hopeless. With this in mind the efficacy of encapsulated antimicrobials was not examined in our burn study. Rather we examined the local concentration of agent over time, the absorption and the ultimate distribution of the liposomes themselves when applied topically in our burn model.

Several of the same characteristics of liposomes discussed in reference to topical treatment also explain the local tissue persistence demonstrated in burn wounds. First, the large size of liposomes tends to keep them in the extravascular site where initially applied.[11-13] Second, the clearance of liposomes after both systemic and local application occurs by way of the reticuloendothelial system. In the case of extravascularly applied liposomes, this clearance is mainly by way of the lymphatics.[11,2,13] RES clearance occurs as a result of antigens which become embedded in the lipid bilayer of the liposome, making the liposomes antigenically recognizable by reticuloendothelial cells.[9,14] The antigenic liposomes are then phagocytized by reticuloendothelial cells, mostly PMN's and macrophages, and subsequently cleared. In burn wounds there are significant concentration of RES cells even in the absence of invasive infections. This suggests the RES uptake as the probable dominant means of clearance for liposomal agents in burns.

In the burn study described, there was a persistent tobramycin tissue concentration and a high level of radioactivity at the burn site, confirming sustained local concentrations of liposome encapsulated tobramycin. The low serum tobramycin levels, even in the face of significant tissue concentration, indicate that liposomal encapsulation decreased the absorption of aminoglycosides in our burn model. This low serum concentration is even more remarkable since the reported tissue levels of tobramycin may be underestimated due to part of the tissue being processed for radioactivity assessment. The small size of the burn in this model could have lessened the absorption of tobramycin and, therefore, effected the serum tobramycin levels; however, given the large dose of tobramycin (5.5 mg), significant serum level would still be expected even with a small wound.

While the proportion of radioactivity recovered (percent of recovered radioactive tag) from the dressing and burn site changed over time, the total amount of radioactivity at the site remained relatively stable. During the 72-hour postburn period, the amount of radioactivity recovered from the burn dressing was seen to increase from 34.42 + 4.97% to 57.22 + 4.78% (p<0.01). This finding is likely a consequence of the increasingly friable burn tissue sticking to the adhesive Opsite® dressing. Importantly, most of the recovered radioactivity remained at the burn site, indicating that our liposome preparation was not readily absorbed from the application site. With notable exceptions negligible amounts of radioactivity were recovered from other peripheral tissues. One exception was the liver which had accumulated a significant concentration of radiolabeled liposomes at 24, 48 and 72 hours. Hepatic concentration of radiolabeled liposomes was likely the result of RES clearance of the liposomes from the burn site and subsequent sequestration in the liver, the largest RES organ in the rat. While not different from other tissues; the spleen, kidney and fat also had increasing levels of radioactivity recovered over the 72 hour experimental period. In the spleen this was again likely due to reticuloendothelial clearance and sequestration of liposomes. The progressive increase in radiolabeled liposomes in the kidney may be a result of the renal clearance of encapsulated tobramycin. Finally, the progressive rise in radiolabeled liposomes in peripheral fat may indicate that liposome breakdown into component lipids and free fatty acids resulted in incorporation of the free fatty acids into the peripheral adipose pool. While the amount of recovered radiolabeled liposomes from peripheral fat was insignificant in our study, normalization of labeled liposome concentration for total body adipose tissue may have achieved significant values. The deposition of our radiolabel in the fat, which was a constituent of the liposomes, raises questions of safety when liposomes are formulated with constituents other than those proven safe. Use of phospholipds which are not naturally occurring could certainly prolong regulatory approval of liposome carriers for human use. The high level of radiolabeled liposome in the intestine was attributed to the fact that several animals ingested a part of their dressing.

Our burn data show that topical application of aminoglycosides incorporated into liposomes to burn wounds resulted in little systemic absorption and sustained local levels of aminoglycosides. This combined with the potential for reducing the frequent application of agents because of sustained tissue levels has significant implications in the management of burn wounds. Currently a large portion of the care resources in burns result from the at least twice daily dressing changes required as a part of currently practiced burn wound management. If liposomal delivery of agents to the burn could decrease this to once a day, a significant savings of health care dollars, not to mention increased patient comfort, would result.

In addition, since the liposomes are cleared at least in part by the reticuloendothelial system therapeutic advantages could manifest in treating invasive infections in burn wounds. Further adding to the advantages of liposome drug delivery in burns is the biodegradable nature of liposomes which should result in less residue requiring removal from the application site. Also, agents which may currently be underutilized or not used at all because of rapid systemic absorption or toxicity could, with liposomal encapsulation, be used. Finally, the various wound factors and inflammatory mediators which are currently under investigation and may be potentially useful in burn wound management may be delivered more effectively by liposomal carriers. Not only because liposome delivery would prolong the otherwise short half-lives of these agents but also because liposome delivery would concentrate these agents in the macrophage, the modulator of most of the wound healing and inflammatory responses.

Liposome-encapsulated silver sulfadiazine is currently in early multi-center trials for the treatment of full thickness burns. Initial results are encouraging and indicate some enhanced effectiveness.[15] Certainly further studies are needed to fully realize the utility of liposome delivery of drugs to burn wounds which, from an experimental perspective, seems to be an exciting application for local liposomal delivery.

REFERENCES

1. Peyman GA, Charles HC, Liu KR, Khoobehi B, Niesman M. Intravitreal liposome-encapsulated drugs: A preliminary human report. Int Ophthalmol 1988; 12:175-182.
2. Delgado G et al. A phase I/II study of intraperitoneally administered doxorubicin entrapped in cardiolipin liposomes in patients with ovarian cancer. Am J Obstet Gynecol 1989; 160:812-819.
3. Fountain MW, Weiss SJ, Fountain AG, Shen A, Lenk RP. Treatment of Brucella canis and Brucella abortus in vitro and in vivo by stable pleurilamellar vesicles encapsulated aminoglycoside. J Infect Dis 1985; 152:529-536.
4. Price CI, Horton JW, Baxter CR. Enhanced effectiveness of intraperitoneal antibiotics administered via liposomal carrier. Arch Surg 1989; 124:1411-1415.

5. Fountain MW, Dees C, Schulte RP. Intracellular killing of S aureus by liposomes containing aminoglycosides: A comparison of liposome surface charge and liposome entrapment of amikacin, tobramycin, gentamicin and kanamycin. Curr Microbiol 1981; 6:373-376.

6. Amber L. Liposome encapsulation enhances antibiotic activity. Infec Dis Today 1988; 2:26

7. Walker JL, Mason AD. A standard animal burn model. J Trauma 1968; 8:1049-1051.

8. Price CI, Horton JW, Baxter CR. Topical liposomal delivery of antibiotics in soft tissue infection. J Surg Res 1990; 49:174-178.

9. Juliano RL. Liposomes as drug carriers in the therapy of infectious diseases. Horiz Biochem Biophys 1989; 9:249-279.

10. Brockenbrough EC, Moylan JA. Treatment of contaminated surgical wounds with a topical antibiotic: A double-blind study of 240 patients. Am Surg 1969; 35:789-792.

11. Hirano K, Hunt CA. Lymphatic transport of liposome-encapsulated agents: Effects of liposome size following intraperitoneal administration. J Pharm Sci 1985; 17:915-921.

12. Segal AW, Willis EJ, Richmond JE, Slavin G, Black CD, Gregoriadis G. Morphological observations on the cellular and subcellular destination of intravenously administered liposomes. Br J Exp Pathol 1974; 55:320-327.

13. Jackson AJ. Intramuscular absorption and regional lymphatic uptake of liposome-entrapped inulin. Drug Metab Disp 1981; 9(6):535-540.

14. Lasic D. Liposomes. American Scientist 1992; 80:20-31.

15. Fountain Pharmaceuticals. Clinical trials with liposome encapsulated silver sulfadizine in burn wounds.(unpublished data)

CHAPTER 7

LIPOSOMES IN THE TREATMENT OF MALIGNANT DISEASE

The treatment of malignant disease is another area in which locally administered liposomal agents may eventually have a significant positive impact. Liposome technology can be beneficial not only with conventional cancer therapies but also in more experimental treatments such as those with immunomodulators or genetic manipulation. Again the local administration rather than systemic therapy with liposomal agents may be a more effective method of treatment with either conventional chemotherapeutics or the newer modalities. Moreover the simplified requirements of local administration allow for simplified carrier design while still resulting in more target specific therapy.

Traditional chemotherapeutic treatment regimes used in cancer therapy often have severe dose-related side effects and toxicities and very small therapeutic windows. Liposome delivery of these toxic agents can widen the therapeutic dosage range without increasing the toxicity or side effect profile. Sells et al has demonstrated in human subjects a significant reduction in systemic side effects with liposome doxorubicin given intravenously. Not only did nausea and vomiting decrease but there was much less myelosuppression, manifested as adequate patient WBC and platelet count, than would be expected with similarly administered free drug. Also the frequent side effect of phlebitis associated with free drug was not present with the liposome doxorubicin. In fact repeat dosages of liposomal drug were given through the same veins, something not possible with free drug. This decreased tissue reaction may have further clinical import since extravasation of chemotherapy agents can result in difficult to manage local wounds. In patients with metastatic breast cancer Treat et al have shown that liposome doxorubicin also resulted in decreased myelosuppression and GI toxicity along with a significant reduction in cardiotoxicity. This was while still retaining tumoricidal activity.[1] They also corroborated an absence of phlebitis and venous sclerosis by Sells. They speculate that the reduction in tissue toxicity may allow for local application of liposome chemotherapy agents. This possibility is further explored by Delgado and many of the same authors in a clinical study using intraperitoneal doxorubicin in advanced ovarian cancer. Since ovarian cancer often spreads to the peritoneal cavity resulting in difficult to control disease local chemotherapy is sometimes used. Intraperitoneal chemotherapy often results in dose limiting sterile peritonitis. Delgado demonstrated that intraperitoneal liposome doxorubicin was tolerated better that free drug allowing for increased dosages. Also, in contrast to systemically administered liposome doxorubicin, they found no myelosuppression. From their pharmacodynamic data it is clear that liposomes inhibited the absorption of drug from the peritoneal cavity.

The pharmacodynamic changes resulting from liposomal encapsulation can translate into therapeutic advantages. With free doxorubicin cell entry is mainly by diffusion and penetration is seldom beyond seven layers of cells in tumors. Liposome drug, on the other hand, are actively taken up by endocytosis resulting in much deeper penetration.[2] Also since there is less tissue irritation and systemic absorption liposome drug can be left in the peritoneum allowing for more prolonged contact between tumor cells and drug. The importance of this is that tumorcidal activity is time- and dose-dependent. The CNS treatment of malignancy was explored by Hong and Mayhew.[3] They demonstrated that in mice with a monoclonal CNS leukemia liposome-encapsulated lip-ara-C administered intracranially was significantly superior treatment than either free drug given similarly or liposome drug given intravenously. Other interesting work by Egerdie et al showed that liposome encapsulated cisplatin, thiopenta and methotrexate all were more effective in killing transitional cell tumors (TCCa) in tissue culture than free drug.[4] This may have clinical relevance since intravesicular chemotherapy is used in urology for treating some early malignancies of uroendothelium. Aside from the increased efficacy demonstrated by Egerdie the decrease in absorption of liposomal agents from application sites along with the decrease in local tissue reaction could make it possible to instill intravesicular liposome chemotherapy and leave it for a longer time further enhancing clinical responses. Alone or as adjuncts to systemic chemotherapy liposome agents may find many other applications in the treatment of primary malignant tumors.

Many times the treatment of the primary site of malignancy, either by surgery or chemotherapy, is successful only to ultimately have a patient return with distant or local regional disease. Being able to concentrate anti-tumor agents in the sites of these recurrences theoretically could change the outcome of many patients. Liposome delivery by nature of its RES sequestration may be useful in the application of agents to prevent or treat this microscopic or macroscopic disease. Potter and associates showed that liposomes were more readily taken up by melanoma cells when compared to hepatocytes in culture.[5] Since liposomes are readily taken up and concentrated in RES organs like the liver the preferential uptake by tumor cells offers some interesting possibilities for the treatment and even prevention of hepatic metastatic disease. Hepatic artery catheters are currently used to deliver intraarterial chemotherapy in advanced malignancy. Liposome agents similarly delivered may be more effective and, given the data of Delgado and Sells, less toxic. In the case of GI malignancies, which commonly result in hepatic metastases, intraperitoneal application of liposomal agent, either at the primary surgical treatment or secondarily, could of result in a decrease in the rate of hepatic metastases and also decrease the local recurrence.

The data of Hirano and Hunt certainly supports the hepatic destination of intraperitoneal liposomes. Khato and associates showed that liposome encapsulated melphan was sequestered in regional lymph nodes after subcutaneous administration.[6] They suggest that melphan encapsulated in liposomes could be helpful as adjunctive therapy in breast cancer patients undergoing extripative surgery. These patients often have regional lymph node disease. Since a significant number of patients now undergo limited surgery of the primary tumor and lymph node sampling rather than regional lymphadenectomy this concept of local chemotherapy could save this group of patients the prolonged radiation therapy required with limited surgery. Combining the implications of the study by Khato with that of Potter suggests that local liposome agent may play a role in the primary management of melanoma. Melanomas also tend to spread via regional lymphatics. Application of liposome agents at the site of tumor excision would have the effect of concentrating drug in the same path as tumor spread possibly changing the dismal course of this disease.

Another area in which liposome may have a role is in the treatment of malignancy with immunomodulators. Agents such as muramyl dipeptide (MDP) and interferon are under study as possible therapeutic agents. These agents act via macrophages to stimulate and enhance the immune response. Liposomes result in macrophage sequestration of entrapped agents so they seem ideal as delivery vehicles for immunomodulators. Daemen et al demonstrated that in an experimental model of hepatic metastatic disease liposome encapsulated MDP given intraperitoneally was able to reduce the development of hepatic tumor and with pretreatment they were able to further reduce tumor load.[7] They were also able to correlate the tumor load with the effectiveness of treatment, the lower the tumor load the more effective the immunotherapy. This group hypothesizes that liposome entrapped MDP may have some clinical benefit when given at the time of surgery for patients undergoing primary resective therapy. In this setting micrometastases may result from tumor manipulation so that the protection imparted by liposome MDP may be more clinically relevant. Rutenfranz et al looked at the effects of another immunomodulator, interferon gamma, incorporated into liposomes.[8] This group found that liposome interferon was more effective in the in vitro stimulation of monocytes and natural killer cells. Moreover they found that liposome interferon had more antiproliferative effects than free interferon. The eventual use of liposomes in this setting may be limited however due to the equivocal clinical results these immunomodulators have had to date.

A truly revolutionary area of cancer therapeutics is that of gene therapy. Basically gene therapy involves the insertion of nucleotide sequences, either DNA or RNA, into a cell to modify its characteristics in some favorable way. In nonmalignant disease this may be the induction of an otherwise deficient protein or it may be to modify the proliferative ability of the cell. Gene therapy has been used to successfully treat two patients with the metabolic disease, adenosine deaminse (ADA) deficiency.[9] In malignant disease tumor cells can be induced to produce proteins which make them more antigenically visible. Alternatively the nucleotides could potentially be used to turn off the hyperfunctioning cellular machinery of malignancy.

A major stumbling block to gene therapy use has been the development of safe and effective methods for nucleotide transfer. Physical and chemical methods are generally not practical outside of cell culture. Viral vectors, while effective, generally have overriding safety concerns which do not permit widespread use. Nabel and associates have used DNA encapsulated into liposomes to transfer a beta-galactosidase gene into vascular endothelium and found that the gene was expressed for up to six weeks.[9] This same group is also exploring the malignant therapy potential for liposome DNA coding for the HLA-B7 antigen. Early encouraging laboratory studies have led to the recruitment of human volunteers for clinical trials. In these trials they plan to directly inject liposome DNA into tumor nodules in patients with advanced melanoma. After the DNA is taken up by the malignant cells the HLA-B7 antigen will be expressed acting like a beacon for killer T-cells which will then destroy the tumor cells. In addition, their laboratory studies suggest that the stimulated killer T-cells will also begin to recognize and destroy the unaltered melanoma cells. This group envisions future studies where they inject liposomal DNA into the vascular network of tumors to more completely distribute the nucleotides.[9] The potential universe of applications for this type of technology is far beyond the scope of this monograph. It is clear however that local liposome delivery will be a safe and effective method to implement the clinical uses of this technology.

Local liposomal chemotherapy can result in specific action and decreased toxicities. By application to the tumor site at the time of primary surgical therapy, liposomal agents may play a role in the prevention of metastatic disease. The role of liposomes in the delivery of immunomodulatory agents is less clear mainly due to the uncertain role of these agents in malignant disease therapy. In the transfer of genetic material liposomes will become the preferred method of nucleotide transfer. There is promising potential for local liposome delivery of agents in the treatment of malignant diseases.

REFERENCES

1. Treat J, Greenspan A, Forst D et al. Antitumor activity of liposome encapsulated doxorubicin in advanced breast cancer: Phase II study. J N C I 1990; 82(2):796-710.
2. Delgado G et al. A phase I/II study of intraperitoneally administered doxorubicin entrapped in cardiolipin liposomes in patients with ovarian cancer. Am J Obstet Gynecol 1989; 160(4):812-819.
3. Hong F, Mayhew E. Therapy of central nerviuos system luekemia in mice by liposome entrapped 1-B-D arabinofuransylcytosine. Cancer Res 1989; 49:5097-5107.
4. Egerdie RB, Reid G, Trachtenberg J. The effect of liposome encapsulated antineoplastic agents on transitional cell carcinoma in tissue culture. J Urol 1989; 142:390-398.
5. Potter DA, Gorchboth CM, Schnieder PD. Liposome uptake by melanoma: In vitro comparison with hepatocytes. J Surg Res 1985; 39:157-163.
6. Khato J, de Campo A, Sieber S. Cancer activity of sonicated small liposomes containing melphan to regional lymph nodes of rats. Pharmacology 1983; 26:260-270
7. Daeman T, Dontje BH, Veninga A, Scherphof GL, Oosterhuis WL. Therapy of murine liver metastases by administration of MDP encapsulated in liposomes. Selective Cancer Ther 1990; 6(2):63-71.
8. Rutenfranz I, Bauer A, Kirchner H. Pharmacokinetic study of liposome-encapsulated human interferon-gamma after intravenous and intramuscular injection in mice. J Interferon Res 1990; 10:337-341.
9. Nabel EG, Plautz G, Nabel GJ. Gene transfer into vascular cells. JACC 1991; 17(6):189b-194b.

=========================== CHAPTER 8 ===========================

THE USE OF LIPOSOMES FOR INTRACELLULAR DELIVERY OF FREE RADICAL SCAVENGERS

Numerous studies have shown that toxic oxygen metabolites including superoxide, hydrogen peroxide and the hydroxyl radical, are produced as byproducts of normal cellular metabolism, primarily in the intracellular compartment. These toxic oxygen metabolites are highly reactive but are capable of relatively short diffusion distances. Numerous studies over the last ten years have shown that a number of situations which alter oxygen delivery to tissue, including ischemia, hypoxia, hyperoxia, and several types of trauma significantly enhance the production of the reactive oxygen metabolites. The cellular events which are initiated by any one of the above insults include adenosine triphosphate conversion to hypoxanthine, xanthine oxidase activation, and subsequently an increased production rate of reactive oxygen metabolites. For example, inhalation of increased oxygen concentrations increases production of toxic oxygen metabolites, contributing to alveolar and endothelial cell damage, pulmonary edema, and pulmonary oxygen toxicity.

Whether the production of toxic oxygen metabolites occurs secondary to regional ischemia or secondary to exposure to increased oxygen concentration, the sequela of events which include cell membrane damage and lipid peroxidation appears to be similar. A number of studies have attempted to treat ischemia-mediated and hyperoxia-induced cell membrane damage by administering free radical scavengers to enhance the scavaging capacity of endogenous antioxidant enzymes. However, since toxic oxygen metabolites are capable of relatively short diffusion distances and are produced in the intracellular compartment, enhancement of the cellular natural antioxidant mechanisms must necessarily include free radical scavenger delivery to the specific areas of toxic oxygen metabolite production. Intravenous administration of the antioxidant enzymes superoxide dismutase and catalase has proved relatively unsuccessful in decreasing the lethal side effects of hyperoxic exposure and has shown only limited success in treating ischemia-mediated cell membrane damage, likely due to limited uptake of the enzyme by the damaged cells. Administration of exogenous superoxide dismutase and catalase by intraperitoneal and aerosol routes has frequently failed to modify postischemic injury and has failed to provide protection against pulmonary oxygen toxicity. Reasons for this failure include the fact that the circulation half-life of the antioxidant enzymes is less than 30 minutes, and the rapid clearance of antioxidant enzymes from the blood stream prevents accumulation of the enzymes in specific intracellular sites where free radical production is greatest. Attempts to enhance the endogenous scavenging capacity of the antioxidant enzymes have included increasing circulation half-life of the enzymes by binding superoxide dismutase or catalase to a delivery vehicle such as polyethylene glycol or ficoll; while these efforts have significantly increased the circulation half-life of antioxidant enzymes, several studies have suggested that binding enzymes to these large molecules limits enzyme delivery into the intracellular compartment.

Recent studies have suggested that liposomes are an effective transmembrane delivery system, and provide an effective method for augmenting cellular antioxidant defenses and decreasing free radical mediated cell membrane damage. The studies described in this chapter include adequate control groups as demonstrated by the administration of empty liposomes as well as administration of unbound free radical scavengers (scavengers without liposomal incorporation); in addition, these studies have compared the efficacy of liposomal antioxidant therapy in several models where unbound scavengers have failed to produce effective protection. The studies described in this chapter have utilized several routes of liposomal enzyme administration including intravenous, intratracheal and intraperitoneal, and have confirmed effective protection against cell membrane damage with all routes of administration. Furthermore, one study confirmed that liposomal delivery of free radical scavengers to endothelial cells grown in culture effectively prevented hyperoxia-induced injury.

The development of liposomes as a transmembrane delivery system has provided a novel means for delivering antioxidant enzymes in several difficult-to-treat situations, for example in pulmonary and brain oxygen toxicity which occurs after exposure to prolonged hyperoxia. Several studies have shown that the brain, like the lungs, is highly sensitive to the effects of increased oxygen exposure. Central nervous system oxygen toxicity, generally attributed to the increased production of oxygen metabolites by cellular processes, results in generalized convulsions. Intravenous delivery of antioxidant enzymes has proven ineffective due to the inability of the enzymes to cross the blood brain barrier and concentrate antioxidant therapy at the site of increased free radical production. The use of liposomes as a delivery vehicle for antioxidant enzymes could concentrate normally membrane-impermeable enzymes in the intracellular compartment and effectively protect against cerebral oxygen toxicity. Furthermore, treatment of hyperoxia-induced pulmonary injury by instilling liposome-encapsulated superoxide dismutase and catalase directly into the trachea, directly targets the lung with antioxidant protection and prevents liposome uptake by the liver and spleen.

LIPOSOMAL PROTECTION AGAINST HYPEROXIA INDUCED CEREBRAL DAMAGE

Yusa and colleagues incorporated bovine-derived superoxide dismutase and catalase into liposomes prepared from dipalmitoyl L-alpha-phosphatidylcholine, cholesterol and stearylamine by reverse phase techniques. After incorporating enzymes into the liposomes, soluble superoxide dismutase and catalase were separated from liposomal entrapped enzymes by centrifugation. All liposomes, including empty liposomes which served as controls, were prepared the day before animal experimentation. Enzyme content of the liposomes was analyzed by enzymatic assay following lysis of the liposomal suspension and sonication.[1] Adult Sprague-Dawley rats, weighing 250-300 grams were included in the study described by Yousa and colleagues. Immediately prior to the study, rats were injected with latex beads (0.7 μM in a buffer solution at pH 7.4) to block reticuloendothelial (RES) uptake of liposomes. One hour after RES blockade, 2 ml of liposomal entrapped enzymes were injected by tail vein. Hyperbaric oxygen exposure consisted of placement of rats into a sealed plexiglass box which had the internal environment regulated through a separate flow system. The entire plexiglass box was then placed inside the pressure chamber after the plexiglass box was sealed and flushed with 100% oxygen; pressure in the chamber was then increased to 1 ATA per minute to 6 ATA and the pressure maintained at this level. The animals were followed by two independent observers until the onset of convulsion and followed until progression to generalized convulsion. Animals were then sacrificed and autopsied; the brain was removed and assayed for enzyme activity; brain tissue enzymatic activity was corrected for blood enzyme contamination.

Liposomal delivery of superoxide dismutase and catalase augmented brain enzymatic activity 2.7 fold and 1.9 fold, respectively by 15 minutes after injection of the enzymes. Pretreatment of rats with liposomal entrapped enzymes two hours before exposure to hyperbaric oxygen increased the time required to produce convulsions by three times that observed in controls. (Fig. 1) In addition, superoxide dismutase plus catalase liposomes, injected at time zero and followed 15 minutes to six hours after hyperbaric oxygen exposure, extended the time to convulsion; the greatest protection against convulsion was achieved by antioxidant liposome injection two hours before hyperbaric oxygen exposure. (Fig. 2) Protection against convulsions correlated significantly with increased brain catalase activity.

Increased production of free radicals during hyperbaric oxygen exposure has been attributed to an increased tissue pO_2 (sometimes exceeding 1000 torr); in addition, an increase in tisssue pCO_2 will increase brain perfusion due to autoregulatory phenomenon. Therefore, hyperoxia accompanied by an increased tissue pCO_2 would be expected to: 1) increase tissue pO_2 secondary to the hyperoxia alone and; 2) increase tissue pO_2 values secondary to CO_2 mediated vasodilatation and an increase in arterial blood flow to the brain. It is well recognized that free radical production during hyperbaric oxygen

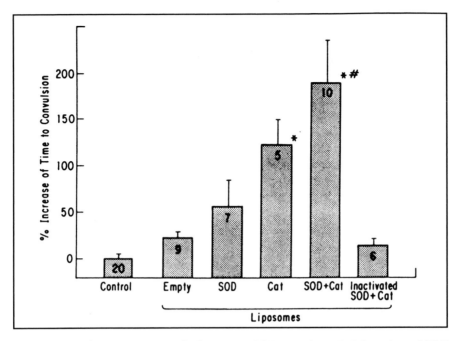

Fig. 1. Effect of empty, superoxide dismutase (SOD), catalase (Cat), inactivated SOD plus catalase, or enzymatically active SOD plus catalase liposomes on time to convulsion in rats exposed to 7 ATA 100% O_2 Rats were injected intravenously with 2 ml liposomes and 2 h later exposed to 6 ATA 100% O_2 Enzyme activity of liposome suspensions was 2,55-6,800 U superoxide dismutase/ml and 31,000-94,000 U catalase/ml. Values are expressed as percent increase of time toconvulsion with control being 24.66± 1.44 min. Data represents mean± SE. *P< 0.05 compared with control (no. treatment with liposomes) and empty liposomes. #P < 0.05 compared with SOD liposomes.

exposure leads to lipid peroxidation of brain cell membranes, altered brain electrical activity, and eventual convulsions.[1] In addition, toxic oxygen metabolites have been shown to alter sodium potassium-ATPase activity, alter intracellular enzymatic activity, disturb energy metabolism, inhibit glutamate decarboxylase, all potential elements in the development of oxygen toxicity in the brain. The incorporation of various substances such as superoxide dismutase, catalase, beta-galactosidase, or gamma aminobutyric acid into liposomes increased brain uptake due to liposomal-mediated passage across the blood brain

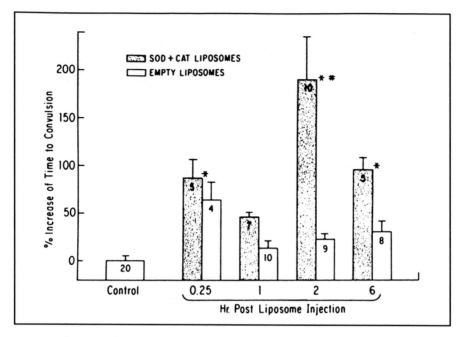

*Fig. 2. Influence of postinjection time of superoxide dismutase (SOD) plus catalase (CAT) and empty liposomes on time to convulsion in rats exposed to 6 ATA 100% O₂. Rats were injected intravenously with 2 ml empty or SOD plus catalase liposomes (2,500-6,800 U superoxide dismutase/ml and 31,000-94,000 U catalase/ml) then exposed to 6 ATA 100% O₂ at indicated times after intravenous liposome injection. Values are expressed as percent increase of time to convulsion compared with control (24.66±1.44 min) and as means ±SE with n in bar. *P<0.05 compared with control (no. treatment with liposomes). #Significantly different from empty liposome and SOD liposome groups.*

barrier. [2] Incorporation of pharmaceutical agents into liposomes ensured delivery across the blood brain barrier due to the ability of the liposomes to fuse with membranes, subsequently releasing their entrapped agents by exocytosis. These data are consistent with reports by Loeb and colleagues[9] who described that incorporation of gamma aminobutyric acid into liposomes and administered intraperitoneally inhibited penicillin-mediated seizures in rats while the intraperitoneal administration of free gamma aminobutyric acid failed to produce an antiepileptic effect.[3]

LIPOSOMAL DELIVERY OF ANTIOXIDANT ENZYMES IN TREATING PULMONARY DYSFUNCTION

Other studies have examined the efficacy of liposomal encapsulation as a delivery vehicle for antioxidant enzymes to prevent pulmonary toxicity after exposure to elevated levels of inspired oxygen. In one study by Padmanabhan and colleagues, superoxide dismutase and catalase were encapsulated into phosphatidylcholine liposomes, and the liposomes were separated from nontrapped enzymatic enzymes by differential centrifugation.[2] Liposomal entrapped superoxide dismutase and catalase activity were measured after lysing an aliquot of the liposomal suspension in a mixture of Triton-X-100 followed by sonication. The authors described that at least 20% of the enzymes were encapsulated into liposomes using this technique. In their study, adult Sprague-Dawley rats, weighing 300-350 grams, were pretreated with intratracheal injection of liposome encapsulated superoxide dismutase and catalase prior to hypoxic exposure. Experimental groups included liposomal enzyme treatment, treatment with free enzyme, and treatment with empty liposomes. The animals were placed in a polystyrene chamber equipped with an alkaline CO_2 absorber, and hyperoxic exposure was continuous at 23-25°C (chamber infused with 100% oxygen at 15 liters per minute). Analysis of the chamber contents showed a pO_2 greater than 715 mmHg and a pCO_2 less than 10 mmHg. Surviving animals were sacrificed at the end of 24 and 72 hours of hyperoxic exposure with pentobarbital injection. Lungs from sacrificed animals or animals dying during the experiment were collected, examined for gross appearance of lung hemorrhage and then perfused with ice cold phosphate buffer, dried, weighed and homogenized with a polytron tissue homogenizer. After differential centrifugation, the supernatants were analyzed for SOD and catalase enzymatic activity. Pretreatment with liposome encapsulated SOD and catalase significantly increased enzymatic activity in lung homogenates compared with lung homogenates from control rats; increased enzymatic activity was apparent at both 24 and 72 hours of hyperoxic exposure. Increased survival rate after 72 hours of hyperoxic exposure correlated with increased pulmonary concentrations of superoxide dismutase and catalase; and all SOD and catalase treated rats had less lung injury than did their respective control groups.

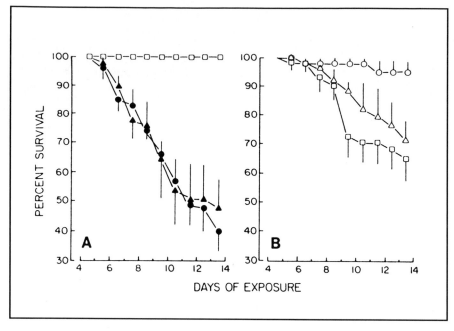

Fig. 3. A. Survival (%) of newborn rat pups exposed to air (□) or >95% O₂ with (●) or without (▲) daily intraperitoneal injections of phosphate buffered saline, for 13.5 days from birth. n=5 litters for each group. B. Survival (%) of rat pups exposed to >95% O₂ for 13.5 days from birth. Pups received daily intraperitoneal injections of either control liposomes containing no enzymes (△) liposomes with entrapped SOD and catalase (●), or liposomes containing active SOD but inactive catalase (□). n=5 litters for each group.

Similar studies by Tanswell and Freeman[3] examined the feasibility of treating newborn rats with liposome incorporated superoxide dismutase and catalase to prevent pulmonary oxygen toxicity after hyperoxic exposure. Timed-pregnant Sprague-Dawley rats were placed in the hyperoxic chamber approximately four hours prior to the anticipated time of delivery. Initial studies determined the lethal time for 50% of the population of newborn pups, a time when death could be attributed to hyperoxic exposure. Subsequent studies included daily intraperitoneal injection of liposome entrapped superoxide dismutase and catalase or vehicle alone. Intraperitoneal injection of liposomal entrapped enzymes were begun 12 hours after birth and repeated at 24 hour intervals until completion of hyperoxic exposure. Liposomes

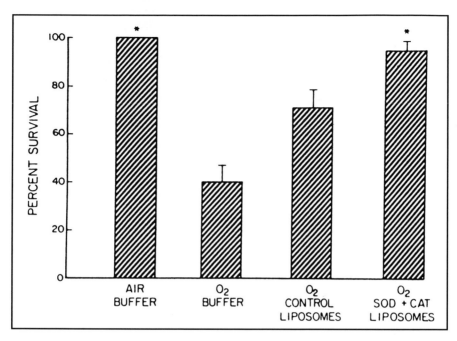

Fig. 4. Survival (%) after 13.5 days exposure to air or > 95% O_2 with daily intraperitoneal injections of liposome suspension buffer, or 95% O_2 with injections of control liposomes or liposomes containing both SOD and catalase. n = 5 litters for each group. All bars having the same symbol are not significantly different from each other.

were prepared from a phospholipid, cholesterol and stearylamine mixture in a phosphate buffer; after liposomal preparation, SOD and catalase were incorporated into the liposomal preparation and an assay of SOD and catalase activity was performed as previously described.[1] In the study by Tanswell and Freeman, daily administration of liposomal entrapped antioxidant enzymes increased survival after 13.5 days of hyperoxic exposure from 40 ± 7% of neonatal rats treated with buffered saline alone to 95 ± 4% with SOD-catalase treatment. (Fig. 3) Injection of empty liposomes (contained no antioxidant enzymes) increased survival from 40 ± 7 to 71 ± 7 days after hyperoxic exposure. (Fig. 4) The investigators concluded that some protective effects were achieved with the administration of empty liposomes; however, the protective effect was greatly exaggerated when antioxidant enzymes were incorporated into the liposomal preparation (p=0.05). This study in neonatal rats confirmed the authors previous work in adult rats showing that liposomal SOD and catalase protect against hyperoxia-induced pulmonary toxicity. They attributed the protective effect of empty liposomes to a dietary lipid supplement and improved caloric intake. In this regard, several studies have shown that any enhancement in saturated fatty acid content reduces peroxidation of unsaturated fatty acids and limits hyperoxia-mediated oxidative cell membrane damage.[4]

A subsequent study by Turrens and colleagues[5] examined the effectiveness of liposomal entrapped superoxide dismutase and catalase in providing protection against oxygen toxicity; these studies were performed in adult rats and used intravenous injection of the liposomes. Liposomes were prepared as previously described; bovine liver catalase and bovine superoxide dismutase were incorporated into liposomes as described by Yusa and colleagues.[1] As described with previous studies, adult male Sprague-Dawley rats, weighing 325-350 grams were injected via the tail vein with the liposomal preparation; animals were killed at various time intervals after intravenous liposome injection to study the kinetics of liposomal clearance by the lungs. Subsequent studies evaluated liposomal entrapped antioxidant enzyme protection; rats were pretreated via the tail vein with latex beads to block the RES system; either saline, liposomal entrapped SOD and catalase, or free enzymes were injected one hour after latex bead treatment, and the animals were immediately exposed to 100% oxygen as previously described.[1] Liposome encapsulated enzymes were injected every 12 hours of hyperoxic exposure until the 36th hour. Liposome incorporated catalase and superoxide dismutase significantly improved survival (Fig. 5) and reduced fluid accumulation in the pleural cavity after hyperoxic exposure as well as increasing lung wet weight at autopsy; lung superoxide dismutase and catalase enzymatic activity was increased 1.7 and 3.1 fold, respectively, two hours after injection of liposomal enzyme. In contrast, injection of free superoxide dismutase and catalase had no effect on survival time, and there was no increase in pulmonary enzymatic activity in the group treated with free enzymes. Furthermore, injection of free enzymes did not alter pleural effusion volume or lung wet weight after hyperoxic exposure. Previous studies have shown that the half life of free SOD or catalase in the circulation is 6 and 23 minutes respectively (Table 1); in this study by Turrens and colleagues, liposomal encapsulation of superoxide dismutase and catalase increased circulation half life to 4 hours and 2.5 hours respectively. (Figs. 6 and 7)

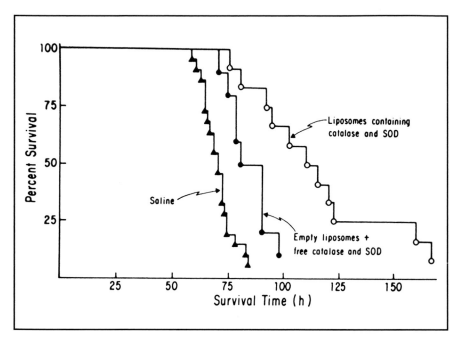

Fig. 5. Effect of liposomal treatment on survival of rats in 100% oxygen. Animals were injected intravenously every 12 h with saline (▲), empty liposomes (equivalent to 85 μmol DPPC) plus 3 x 10³ U SOD and 9 x 10⁴ U catalase (●) or liposomes containing ~3 x 10³ U SOD and 9 x 10⁴ U catalase (○).

Another study by Freeman and colleagues[4] examined the efficiency with which liposomal entrapped superoxide dismutase could be delivered intracellularly to aortic endothelial cells cultured in vitro. These investigators hypothesized that an increase in intracellular superoxide dismutase activity would protect endothelial cells from cellular damage after hyperoxic exposure. Liposomes were prepared as described Yusa and colleagues; human superoxide dismutase purified from liver according to McCord and colleagues,[6] bovine superoxide dismutase, porcine superoxide dismutase or inactivated bovine superoxide dismutase were incorporated into liposomal preparations.[1] Porcine aortic endothelial cells were grown in culture, and cells were incubated for two hours with liposomes suspended in the media. After washing with Hanks solution, cells were reincubated in selected oxygen tensions; plates were maintained in an incubator and exposed to a variety of experimental conditions,

Table 1. Lung Wet Weight, and Pleural Effusion Volume of Animals

Treatment*	n	Lung wet weight	n	Pleural effusion	n	Survival time
		g		ml		h
Saline	27	2.92±0.16‡	24	9.8±0.4‡	21	69.5±1.5‡
Free catalase plus SOD	8	2.66±0.09‡	8	9.2±1.2‡	11	71.4±1.7‡
Empty liposomes	8	2.98±0.18‡§	8	9.4±0.7‡	10	71.2±1.5‡
Catalase liposomes	7	3.57±0.43‡§	6	4.9±1.4	7	74.4+7.3‡
SOD liposomes	7	2.80±0.16‡§	7	9.0±0.5‡	7	72.4±1.6‡
Empty liposomes with free SOD, SOD plus catalase	9	3.80±0.25§	10	2.1±1.0§	9	83.1±3.1‡
Liposomes containing SOD plus catalase	11	4.83±0.36	11	0.5±0.3§	12	118.1±9.9

*Rats were injected intravenously with a total volume of 2 ml saline or a suspension of liposomes (60-90 µmol DPPC) containing either saline, catalase (8-9.8 x 10^4 U), or SOD (2-3 x 10^3 U) suspended in saline. Free catalase (5 x 10^4 U/ml) and SOD (2 x 10^3 U/ml) were injected in 2 ml saline.
‡§Values having the same symbol superscript are not significantly different from each other (P<0.05).
All values represent mean ± SEM and are compiled from experiments done on three to five separate occasions.

including 5% oxygen/95% air, room air, or 95% oxygen/5% room air; thus, oxygen tension was maintained at 5, 21, or 95%. Cytotoxicity was assessed by the 51-chromium or lactate dehydrogenase release from endothelial cell into the culture media. Previous studies have shown a close correlation between chromium release and lactate dehydrogenase release after oxidant damage in tissue slices.[7] Exposure of endothelial cells to hyperoxia increased 51-chromium and lactic dehydrogenase release in a manner that correlated significantly with the degree of hyperoxia. Liposomal entrapment of superoxide dismutase increased cellular antioxidant enzymatic activity 6-12 fold. Liposomal mediated augmentation of superoxide dismutase concentration in endothelial cells increased resistance to oxygen damage compared to that observed

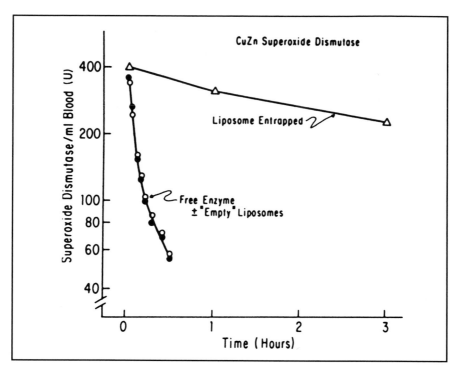

Fig. 6. Clearance of free and lipoosme-entrapped CuZn SOD from the rat circulation. Rats were injected in the tail vein with enzyme and/or liposomes having a total volume of 1.0 ml, in 0.15M NaCl, 10 mM potassium phosphate, pH 7.4. Injections consisted of 4,000 U CuZn SOD (●), 4,000 U CuZn SOD plus "empty" liposomes containing 40 μmol phospholipid (○), or 4,000 U CuZn SOD entrapped in liposomes containing 40 μmol phospholipid (△). At selected intervals, 100-μl blood samples for enzyme assay were removed from the inferior vena cava of pentobarbital-anesthetized rats, centrifuged, hematocrit determined, and enzyme activity of the serum determined. Activity per milliliter whole blood was then calculated from the packed cell volume of each sample.

in the untreated controls, and the decrease in cellular damage correlated significantly with the endothelial concentration of antioxidant enzymes. Addition of free superoxide dismutase to the endothelial cells in culture did not increase cellular enzymatic activity and did not prevent oxygen-induced injury. Addition of empty liposomes (liposomes prepared as described above but containing no superoxide dismutase) had a small and significant protective effect from hyperoxia mediated cell damage, likely due to a liposome-mediated increase in cell membrane lipid content and a decrease in cell membrane peroxidation.

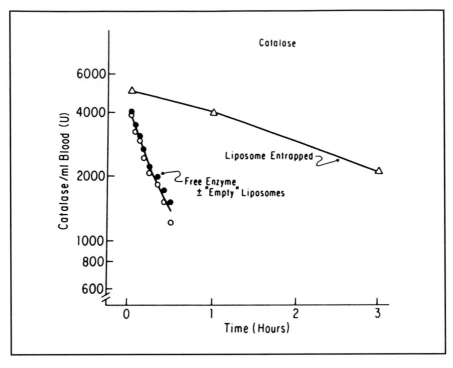

Fig. 7. Clearance of free and liposome-entrapped catalase from the rat circulation. This experiment was performed as in Fig. 6, except rats were each injected with 60,000 U catalase. (●) represent rats injected with free catalase; (○) represent free catalase plus "empty" liposomes; and (▲) represent free catalase plus "empty" liposomes; and (△) represent rats injected with 60,000 U catalase entrapped in liposomes containing 40μmol phospholipid.

LIPOSOMAL DELIVERY OF ANTIOXIDANT ENZYMES IN COLD-MEDIATED AND ISCHEMIA-INDUCED BRAIN INJURY MODEL

A study by Chan and colleagues used liposome-entrapped superoxide dismutase in a cold-induced brain injury model; in their rat model, injury was characterized by a loss of blood-brain-barrier permeability, brain edema, and focal brain necrosis.[10] In their study, liposomes were prepared from dipalmitoyl phosphatidylcholine, cholesterol and sterylamine using reverse phase evaporation.[11] The incorporation of the CuZn-SOD into liposomes yielded 25-40% of antioxidant enzyme incorporation into the liposomes; free antioxidant enzyme was removed from the preparation using standard techniques. In their study, a freezing lesion was produced on one area of the cortex and the contralateral side served as control. The concentration of CuZn-SOD in brain tissue was measured in tissue homogenates by the ability to inhibit superoxide radical dependent reactions. Blood-brain barrier permeability was determined by using an Evans blue technique, and confirmed by intravenous injection of ^{125}I-bovine serum albumin followed by calculating the percentage of ^{125}I-BSA in the brain divided by the units per milliliter of plasma; water content was determined from dry/wet weight ratios.

This study demonstrated that cold injury produced a seven-fold increase in blood-brain barrier permeability as indicated by an altered ^{125}I-BSA space one hour after injury; cold injury-induced changes in blood-brain barrier permeability were confirmed by significant leakage of Evans blue. Intravenous injection of liposome entrapped CuZn-SOD, either five minutes before or five minutes after cold-induced injury, significantly reduced blood-brain permeability and ameliorated brain edema; in addition, antioxidant enzyme therapy reduced the level of superoxide radicals documented in the brain after untreated cold-induced injury.

Chan and colleagues hypothesized that changes in blood-brain barrier permeability and brain edema which occur after cold-induced injury are related to xanthine oxidase mediated formation of oxygen derived free radicals. Xanthine oxidase activity is high in the capillaries of the blood-brain barrier compartment, suggesting that superoxide radicals are formed within intracellular compartment after cold-induced activation of xanthine oxidase. The authors further hypothesized that cerebral tissue membrane phospholipids contribute to the formation of arachidonic acid, and subsequently to superoxide radical formation. The data from this study support their previous finding that a loss of blood-brain barrier permeability occurs after intracerebral injection of arachidonic acid or xanthine oxidase. Their data confirm that superoxide radicals contribute, in part, to brain edema in cold trauma induced injury. In their study, incorporation of antioxidant enzymes into liposomes provided an effective means for enzyme delivery at the site of cerebral free radical production.

A subsequent study by Chan and colleagues[11] examined the effects of liposome encapsulated superoxide dismutase delivery in preventing brain edema and cell membrane damage after focal cerebral ischemia in a rat model. In this model, ischemia was produced by occlusion of the middle cerebral artery and liposome entrapped CuZn-SOD, prepared by reverse phase evaporation and containing dipalmitoyl phosphatidyl-choline, cholesterol and sterylamine, was administered at the time of the time of ischemic insult. These investigators demonstrated that antioxidant enzymes incorporated into liposomes could be successfully delivered in large concentrations to the brain tissue. A single injection of liposome entrapped CuZn-SOD elevated brain enzyme concentrations more than 150%, and these levels persisted up to two hours after intravenous injection. In contrast, an IV injection of free CuZn-SOD did not appreciably increase brain levels of antioxidant enzyme. In addition, the authors demonstrated that liposome incorporated CuZn-SOD significantly reduced brain edema, while liposomes administered without antioxidant enzyme did not alter ischemia-induced brain edema. The authors conclude that oxygen derived free radicals contribute to the cellular dysfunction which occurs after ischemia.

Antioxidant Therapy in Experimental Liver Necrosis

Another successful application of liposome encapsulated antioxidant enzymes was described by Dai Nakae and colleagues.[12] In their study, adult male rats were treated with 500 mg/kg body weight acetaminophen, a regimen associated with considerable liver cell necrosis as indicated by a significant rise in serum concentration of aspartate aminotransferase, alanine aminotransferase, and alkaline phosphatase. Liposome encapsulated superoxide dismutase was administered by intraperitoneal injection at the time of acetaminophen administration. In their study, liposome encapsulated superoxide dismutase provided significant protection from acetaminophen induced liver necrosis. In contrast, free SOD, heat denatured SOD, or SOD free liposomes exerted no protective effect. The authors also showed that encapsulated SOD did not alter acetaminophen metabolism as indicated by no change in the rate or extent of hepatic glutathione depletion. Lipid peroxidation occurred in their model of acetaminophen induced liver necrosis as indicated by an accumulation of malondialdehyde within three hours after administration of acetaminophen; administration of liposome encapsulated SOD prevented malondialdehyde accumulation. In some animals, liver necrosis was potentiated by pretreatment with 3-methylcholanthrene before acetaminophen administration. In their study, liposome encapsulated SOD also prevent the hepatotoxicity of acetaminophen excerbated by 3-methylcholanthrene. The protective effect of liposome encapsulated SOD was confirmed by histologic examination of the liver showing complete amelioration of any evidence of liver cell necrosis in rats that had received 500 mg/kg body weight acetaminophen plus liposome encapsulated SOD (570 units of SOD/mg of liposomal lipid). The authors conclude that liposome encapsulated superoxide dismutase prevented the induction of liver necrosis and attributed hepatotoxicity of acetaminophen to superoxide radical formation. Their finding that free enzyme failed to prevent acetaminophen induced liver necrosis was attributed to: 1) the short circulation half-life of the enzymes after intravenous administration and; 2) the fact that intravascular administration of free enzymes fails to concentrate the antioxidant enzymes at specific intracellular sites where they are required.

A subsequent study that applied liposome technology to the delivery of free radical scavengers was described by Dai Nakae and colleagues[13] who compared the effectiveness of liposome encapsulated superoxide dismutase to that observed with application of free superoxide dismutase in reducing the toxic effect of tert-butyl hydroperoxide on hepatocytes grown in culture. The authors had previously observed that liposome encapsulated antioxidant enzyme prevented liver necrosis induced by acetaminophen while free enzyme was ineffective; however, they were interested in comparing the protective effects of liposome versus free antioxidant enzymes from a quantitative point of view. In this subsequent study, hepatocytes were isolated from adult male Sprague-Dawley rats using collagenase in accordance with previously established protocols. After establishing viabilities of the hepatocytes grown in culture, tert-butyl hydroperoxide (TBHP) was added to the culture at a final concentration of 500 μM; the toxicity of this agent for rat hepatocytes had been established previously. Human CuZn superoxide dismutase was incorporated into a liposomal suspension as previously described; and the resulting liposomes contained 571 units of SOD per mg of liposomal lipid. The authors compared the addition of liposome encapsulated SOD, empty liposomes, free enzyme, and liposomes containing heat inactivated SOD added simultaneously with TBHP to hepatocytes in culture; SOD activity as well as the killing capacity were assessed.

The authors reported that in the absence of SOD, TBHP killed 69% of the cells within one hour. Simultaneous addition of liposome encapsulated SOD to the culture medium reduced the cell killing by 50%; the concentration of liposomal SOD which produced this decrease in cell killing was 32 units per ml and maximal protection was achieved with 64 units per ml. In contrast, significantly higher concentrations of free SOD were needed to achieve a similar reduction in cell killing; 288 units per ml, a ten fold greater concentration of free SOD compared to liposomal incorporated SOD, was required to reduce the toxicity of TBHP by 50%; maximal protection was achieved with 500 units per ml. Neither SOD free liposomes nor liposomes which contained heat denatured SOD were effective at reducing TBHP-induced cell killing. Measurements of cellular SOD content confirmed that liposome encapsulation significantly facilitated the hepatocyte uptake of SOD. The authors concluded that liposomal encapsulation of superoxide dismutase was singularly effective in reducing cell killing by TBHP, likely due to the more efficient delivery of the antioxidant enzymes to the intracellular compartment of the hepatocytes. These studies are consistent with previous reports that liposomal encapsulation of superoxide dismutase facilitates intracellular accumulation of the antioxidant enzyme.[14] Furthermore, the authors confirmed that liposomal delivery of antioxidant enzymes was independent of the endocytosis which was demonstrated as the method of uptake of free antioxidant enzymes.

In a similar study, Tanswell and colleagues[15] examined the response of fetal rat lung fibroblast to elevated oxygen concentrations after treatment with liposomal encapsulated antioxidant enzymes. Fetal lung fibroblasts were grown to confluence in Dulbecco MEM:Ham F12 medium. The cells were plated in culture flasks and gassed with 1% oxygen, a concentration similar to oxygen exposure for fetal tissues in utero. After 24 hours, gas and media were changed and flasks were placed in 1%, 21%, 50% or 95% oxygen. Cell counts were performed on flasks from each group on days five and seven. Fibroblast DNA synthesis was measured and culture medium was assayed for LDH, d-lactate, d-hydrogenase and prostaglandin content.

Superoxide dismutase and catalase were added to the liposomal preparation as previously described, and unencapsulated enzyme was removed by centrifugation. The authors have shown that 40% of the antioxidant enzymes in the original solution are incorporated into the liposomes using their preparation techniques. Experimental groups included cell culture plus: 1) liposome suspension buffer, 2) free liposomes, 3) liposome encapsulated SOD, 4) free antioxidant enzymes without liposomal encapsulation. Growth rates of fetal lung fibroblasts were significantly reduced in the presence of 50 or 95% oxygen, and the higher oxygen concentration had the greatest reduction in growth rate. Fetal rat lung fibroblast were growth inhibited, and LDH release and prostaglandin synthesis were increased in the presence of 50 and 95% oxygen exposure. Pretreatment of the fetal rat lung fibroblasts with liposome encapsulated superoxide dismutase and catalase provided significant protection against the cytotoxic effects of 50 and 95% oxygen as measured by a reduced LDH release.

The authors major concern in this study was that parenteral administration of liposomal encapsulated superoxide dismutase and catalase has been shown to prevent the lethal effects of hyperoxia in adult and neonatal animals; however, previous studies have not determined whether this therapy preserved normal lung function and growth. Therefore, these studies were designed to examine the effects of liposome encapsulated SOD and catalase on fetal lung growth as determined in vitro. Earlier studies had shown that significant protection against lethal hydrogen peroxide injury was achieved in adult rat pneumocytes treated with liposomal encapsulated antioxidant enzyme; however, the studies provided no information with regard to cell growth since adult pneumocytes do not divide in culture. Increased intracellular SOD and catalase activity after liposomal delivery protected fetal lung fibroblasts against the lethal effects of 95% oxygen; while this therapy provided a significant measure of protection, it did not prevent hyperoxia induced stimulation of prostaglandin synthesis nor did it prevent inhibition of DNA synthesis. The authors concluded that hyperoxia increases the intracellular production of toxic oxygen metabolites, increasing oxidation of membranes. While their studies confirmed that liposomal delivery enhanced intracellular concentration of antioxidant enzymes and reduced the lethality of hyperoxia, these factors did not prevent hyperoxia induced changes in prostaglandin synthesis. The authors conclude that other scavengers such as glutathione peroxidase may play a greater role in reducing cellular toxicity during hyperoxia.

Summary of Liposomal Delivery of Antioxidants

These studies suggest that the liposomal method of pharmacologic delivery ensures access to intracellular sites where toxic oxygen metabolites are produced. Therefore, liposomal delivery of pharmacologic agents provides an effective means for intracellular delivery of large molecules, and all routes of liposomal administration (intravenous, intraperitoneal and intratracheal) increased the specific activity of the antioxidant enzymes three to five fold in lungs and brain and 10 to 25 fold in cultured endothelial cells. The finding by Yusa and colleagues that liposomal delivery of scavengers reached a maximal effective level two hours after liposomal injection suggests that adequate time is required for transfer of liposomal entrapped agents to intracellular sites; these data are consistent with previous studies in this regard. It is well recognized that hyperbaric oxygen increases blood-brain barrier permeability, increasing the rate of uptake of liposomal entrapped agents and therefore increasing the half-life of liposomal incorporated enzymatic activity in brain intracellular spaces.

The findings of the above studies confirm that liposomal mediated delivery of macromolecules to the brain as well as to the lungs provide a unique and useful method of delivery. The ability of this pharmacologic delivery system to achieve adequate levels of drugs or enzymes in hard-to-reach compartments has far-reaching consequences for both research and clinical application. Furthermore, the entrapment of agents such as superoxide dismutase and catalase as well as numerous other free radical scavengers into liposomal preparations increases the circulation half-life of these agents, preventing intravascular breakdown and clearance before the agents can achieve their desired pharmacologic effects. This provides effective augmentation of intracellular antioxidant defenses and provides an effective tool for both preventing oxygen radical mediated cellular damage and for studying the metabolism of oxygen free radicals in the intracellular compartment.

The mechanism by which liposomal enzyme delivery achieved protection against hyperoxia mediated cell membrane dysfunction is likely related to increased scavenging of superoxide, thereby decreasing secondary reactions of the superoxide radical with cell constituents. Numerous studies have shown that superoxide has direct cytotoxic capability and that this oxygen derived free radical plays a primary role in generating secondary reactive oxygen metabolites. The ability to increase intracellular concentrations of superoxide dismutase in organs that are at risk from ischemic damage is particularly applicable in many trauma and burn states where cell membrane dysfunction and the progression to multiorgan failure has been attributed to the production of toxic oxygen metabolites that overwhelm the scavenging capacity of endogenous enzymes.

The ability of liposomes to cross plasma membranes ensures intracellular entry of membrane permeable agents entrapped within the liposomes. Therefore, liposomes can serve as a delivery vehicle for aqueous compartment entrapped agents as well as a delivery method for hydrophobic compounds incorporated into liposomal membranes. The intracellular delivery of liposome agents occurs first by liposomal binding to the plasma membrane; frequently liposomes undergo endocytosis, providing entry into the lysosomes. Liposomes may also fuse with plasma or with lysosome membranes; subsequently, the liposomal contents may enter the intracellular space or the liposomes may be internalized by the cell without the outermost bilayer. The concurrent occurrence of these processes depends on liposome composition, temperature, acidity as well as a number of other factors including electrolyte composition. Several studies included in this review have suggested that liposomes delivered their pharmacologic contents by fusion or endocytosis with the cell membrane as opposed to liposome adherence to the cell membrane and rupture prior to cellular uptake. Whether liposomal entrapped agents enter the intracellular compartment secondary to liposomal binding to the plasma membrane or release of liposomal contents into the cytoplasm after incorporation of liposomal membrane lipid into cell membranes is irrelevant with regard to the effective delivery of free radical scavengers. The finding that the incorporation of free radical scavengers into liposomes resulted in the delivery of a biologically active form of the enzymes further increases the feasibility of this delivery system.

Another way in which liposomal delivery of free radical scavengers may mediate protection against cytotoxic free radicals is incorporation of the liposomes into cell membranes with subsequent modification of the membrane lipid composition. Several studies have shown that increases in the saturated lipid content of cells protects against oxygen induced cell membrane damage, likely due to decreased lipid peroxidation of the membranes. Incorporation of additional lipids into the membrane component or changes in the saturated to unsaturated fatty ratio of membranes may render certain cell populations resistant to peroxidation by free radicals during the ischemia or hyperoxia process.

The studies described above confirm that encapsulation of antioxidant enzymes into lipid vesicles is accomplished rapidly and effectively; furthermore, liposomal delivery of enzymes significantly enhanced intracellular antioxidant capacity, subsequently protecting cells from oxygen toxicity. The lack of immunogenicity, increased circulation half-life, and enhanced cell entry make the liposomal delivery system a useful tool for delivering free radical scavengers in the clinical setting. Liposomal encapsulation protects the antioxidant enzymes from degradation in the peripheral circulation and can specifically target cells that are at risk after ischemic or hyperoxic insult.

REFERENCES

1. Yusa T, Crapo JD, Freeman BA. Liposome-mediated augmentation of brain SOD and catalase inhibits CNSO$_2$ toxicity. J Appl Physiol (Respirat Environ Physiol) 1984; 57:1674-1681.

2. Padmanabhan RV, Gudapaty R, Liener IE, Schwartz BA, Hoidal JR. Protection against pulmonary oxygen toxicity in rats by the intratracheal administration of liposome-encappsulated superoxide dismutase or catalase. Am Rev Respir Dis 1985; 132:164-167.

3. Tanswell AK, Freeman BA. Liposome-entrapped antioxidant enzymes prevent lethal O$_2$ toxicity in the newborn rat. J Appl Physiol 1987; 63:347-352.

4. Freeman BA, Young SL, Crapo JD. Liposome-mediated augmentation of superoxide dismutase in endothelial cells prevents oxygen injury. J Biol Chem 1983; 258:12534-12542.

5. Turrens JF, Crapo JD, Freeman BA. Protection against oxygen toxicity by intravenous injection of liposome-entrapped catalase and superoxide dismutase. J Clin Invest 1984; 73:87-95.

6. McCord JM, Boyle JA, Day ED, Rizzolo LJ, Salin ML, In: Superoxide and Superoxide Dismutases. Michelson AM, McCord JM, Fridovich I, eds. Academic Press Inc, Ltd, London: 1977:129-138.

7. Martin WJ II, Gadek JE, Hunninghake GW, Crystal RG. Oxidant injury of lung parenchymal cells. J Clin Invest 1981; 68:1227-1288.

8. Takada G, Onodera H, Tada K. Delivery of fungal ß-galactosidase to rat brain by means of liposomes. Tohoku J Exp Med 1982; 136:219-229.

9. Loeb C, Benassi E, Besio G, Maffini M, Tanganelli P. Liposome-entrapped GABA modifies behavioral and electrographic changes of penicillin-induced epileptic activity. Neurology 1982; 32:1234-1238.

10. Chan PH, Longar S, Fishman RA. Protective effects of liposome-entrapped superoxide dismutase on posttraumatic brain edema. Ann Neurol 1987; 21:540-547.

11. Chan PH, Fishman RA, Wesley MA, Longar S. Pathogenesis of vasogenic edema in focal cerebral ischemia. Role of superoxide radicals. Adv Neurol 1990; 52:177-183.

12. Nakae D, Yamamoto K, Yoshiji H, Kinugasa T, Maruyama H, Farber JL, Konishi Y. Liposome-encapsulated superoxide dismutase prevents liver necrosis induced by acetaminophen. Am J Pathol 1990; 136:787-795.

13. Nakae D, Yoshiji H, Amanuma T, Kinugasa T, Farber JL, Konishi Y. Endocytosis-independent uptake of liposome-encapsulated superoxide dismutase prevents the killing of cultured hepatocytes by tert-butyl hydroperoxide. Arch Biochem Biophys 1990; 279:315-319.

14. Michelson AM, Puget K. Cell penetration by exogenous superoxide dismutase. Acta Physiol Scand 1980; 492(Suppl):67-80.

15. Tanswell AK, Olson DM, Freeman BA. Response of fetal rat lung fibroblasts to elevated oxygen concentrations after liposome-mediated augmentation of antioxidant enzymes. Biochimica et Biophysica Acta 1990; 1044:269-274.

CHAPTER 9

THE FUTURE OF LOCAL LIPOSOMAL AGENTS

Local therapy with liposomal agents has been overlooked as an application of liposome technology. While some investigators have used liposomes for local delivery of agents, the majority of liposome investigational work has focused on systemic administration. One advantage of the local use of liposomal agents is that some of the complexities of liposomal delivery are removed. There are many problems associated with systemic liposome delivery including the alteration of liposomal characteristics with incorporation of each different drug into the liposomal membrane. These difficulties are minimized by local use of a liposomal system since the important characteristics in local delivery, the local persistence and nonspecific RES clearance, appear less affected by minor changes in membrane characteristics. The less specific requirements of local liposome delivery could also allow a single liposomal formulation to be used as a delivery system for a multitude of different agents, negating the need to tailor a liposomal formulation for each agent and each application.

PROSPECTS FOR WIDESPREAD APPLICATION OF
LIPOSOME-ENCAPSULATED AGENTS TO LOCAL THERAPY

The ultimate widespread use of local liposomal agents will depend on several factors. Foremost will be the regulatory approval for carriers themselves which can then be used to incorporate already approved drugs without going through a long and unnecessary bureaucratic process for each encapsulated agent. Approved use of the carrier itself would provide a strong incentive to further develop liposome technologies. In addition, production of a liposomal preparation must be cost effective to allow application in common diseases; in this regard, many attempts at liposome therapeutics have focused on situations of limited clinical application. These attempts have resulted in expensive formulations required to produce complex liposomes designed with the elegant specificity required for most systemic applications. The necessity to recover the cost of these formulations make it unrealistic to treat, for example, a minor laceration with a liposomal agent. Finally, treatment strategies should be designed to ensure that the liposomal agent is in close proximity to the target area. Our studies indicate that liposomes are very effective when used to concentrate the delivered agent at the application site and not when expected to produce specific distant organ or tissue concentrations. This again ultimately allows for simplified liposomal construction, eliminating the need for expensive "stealth" or targeted liposomes.

There are many areas in clinical medicine which may benefit from local liposomal drug delivery. The experimental work outlined in this text would suggest that liposome antibiotic delivery may prove to have therapeutic superiority in the treatment of peritoneal and subcutaneous bacterial infections when compared to free drug treatment. In the treatment of superficial and deep burn wounds, liposomal delivery may result in favorable dosage scheduling due to the local persistence of delivered agents.[1-4] As described in this monograph, preliminary clinical studies indicate enhanced effectiveness in the treatment of full thickness burns with the use of liposome encapsulated silver sulfadiazine.[5]

Cancer therapy is also an area where local liposomal delivery has potential widespread utility. In the treatment of malignancy, the liposomal delivery of chemotherapeutic agents may have advantages especially in conditions where there is a high propensity for local recurrence or early lymphatic spread such as malignant melanoma and breast cancer. The local application of liposomal agent and the subsequent lymphatic clearance may help to decrease early recurrences postoperatively especially in the regional lymphatics. This could be particularly important in the treatment of malignant melanoma where regional lymph node disease seems to play such a pivotal role in disease spread. Liposomal delivery of chemotherapy agents in treating cutaneous malignancies may also prove to be a more effective treatment while resulting in less local and systemic toxicities. Liposomes offer advantages not only in delivery of chemotherapeutic agents but also the metabolic and inflammatory modulators that can be used to "turn off" tumor cells, including wound factors or nucleotide fragments which modulate tumor cell growth. With nucleotide fragments, the ability to deliver agents intracellularly with liposomal encapsulation may be pivotal in allowing the clinical application of this technology.

Liposome delivery may also be advantageous in the application of wound modulating agents, most of which have very short tissue half-lives. These wound factors are currently under investigation as a means of reducing scar formation, altering free radical formation, speeding wound closure and modulating the wound healing process.

The use of liposomes in the delivery of free radical scavengers may extend the use of these experimentally useful agents to clinical situations. To date the use of free radical scavengers has proven clinically difficult and unpredictable due to the short circulating half-life of the agents and the difficulty in achieving adequate concentrations of agents in the target tissues. Local application of lipsomal agents eliminates many of these problems.

The concept of liposomal delivery of an agent by direct application to the target tissue has vast clinical potential. The general loss of enthusiasm for liposomal pharmaceuticals has, in part, stemmed from the tunnel vision imposed by systemic administration. Hopefully, this monograph will help rekindle interest in liposome drug delivery such that its ultimate enormous potential and benefits can be realized.

References

1. Price CI, Horton JW, Baxter CR. Enhanced effectiveness of intraperitoneal antibiotics administered via liposomal carrier. Arch Surg 1989; 124:1411-1415.
2. Price CI, Horton JW, Baxter CR. Topical liposomal delivery of antibiotics in soft tissue infection. J Surg Res 1990; 49:174-178.
3. Price CI, Horton JW, Baxter CR. Liposomes: A new method for the delivery of burn wound antimicrobial therapy. SG&O; accepted for publication.
4. Price CI, Horton JW, Baxter CR. Liposome encapsulation: A method for enhancing local antibiotics. Submitted for publication.
5. Fountain Pharmaceuticals. Clinical trials with liposome encapsulated silver sulfadizine in burn wounds.(unpublished data)

INDEX